NATIONAL IDENTITY AND GEOPOLITICAL VISIONS

After a period in which nationalism was considered to be in decline, events in Eastern Europe have forced the world to look afresh at national impulses. It is now clear that one cannot always put the blame on an evil political system and its leaders. Some forms of distrust or hate towards what is foreign appear to be of a genuinely popular nature. What features in the environment and history of a national group arouse such feelings towards the outside world?

National Identity and Geopolitical Visions searches for national orientations in the relationship of a people with the world, a relationship based on the desire for state security and for an influence outside that state.

Through nine country-specific essays – on Germany, Britain, the United States, Argentina, Australia, Russia, Serbia, Iraq and India – the author explores whether there is continuity in national values and foreign policy, and how such geopolitical visions are shaped by national and international events. The pattern is diverse, but geopolitical visions are never the rational evaluation of a country's strategic advantages that the word "geopolitics" suggests.

Gertjan Dijkink is Associate Professor of Political Geography at the University of Amsterdam.

NATIONAL IDENTITY AND GEOPOLITICAL VISIONS

Maps of pride and pain

GERTJAN DIJKINK

LONDON AND NEW YORK

First published 1996
by Routledge
11 New Fetter Lane, London EC4P 4EE

Transferred to Digital Printing 2003

Simultaneously published in the USA and Canada
by Routledge
29 West 35th Street, New York, NY 10001

© 1996 Gertjan Ditkink

Typeset in Photina by Keystroke, Jacaranda Lodge, Wolverhampton

British Library Cataloguing in Publication Data
A catalogue record for this book is available from the British Library

Library of Congress Cataloguing in Publication Data
Dijkink, Gertjan
National Identity and Geopolitical Visions : Maps of pride
and pain / Gertjan Dijkink.
p.   cm.
Includes bibliographical references and index.
1. Nationalism. 2. Geopolitics. I. Title.
JC311.D53 1996
320.5′4′0904—dc20   96–7557

ISBN 0–415–13934–1 (hbk)
ISBN 0–415–13935–X (pbk)

Printed and bound by Antony Rowe Ltd, Eastbourne

CONTENTS

List of figures vii
Preface ix

1 The National Experience of Place 1

2 The Country of *Angst*
(*Germany*) 17

3 Absent Because of Empire
(*Britain*) 36

4 The March of Civilization: Destiny and Doubts
(*United States*) 49

5 The Last Frontier
(*United States*) 59

6 Peripheral Dignity and Pain
(*Argentina*) 72

7 Wandering in Circles
(*Australia*) 86

8 The Eurasian Dilemma
(*Russia*) 95

9 The Empire of Revenge
(*Serbia*) 109

10 Totally Lost?
(*Iraq*) 119

11 A World in Itself
(*India*) 128

12 Conclusion 139

Notes 148
Bibliography 164
Index 175

FIGURES

1 How the French journal *Hérodote* pictures the geographical visions of the Cuban revolutionaries 13

2 'At last! Our dream is coming true.' Britain and continental dictators (1987) 47

3 The difference between the hemispheres (John T. McCutcheon, 1920) 55

4 America and war-mad Europe (Carey Orr, 1933) 56

5 America in future wars: themes of 108 future war novels (percentages) written between 1946 and 1983 68

6 'A house built on co-operation' ('Catholicize!' 'No, circumcize!') 113

7 A Catholic prelate with the U-symbol of the Ustasha, telling his rosary of plucked-out eyes 114

8 Saddam Hussein and Nebuchadnezzar 122

9 Saddam Hussein as benefactor to the Arab world 124

TABLE

1 War trauma in the United States (number of battle deaths) 61

PREFACE

Sometimes generations take long steps in time. The joyous celebrations of the end of the German occupation in the Netherlands were the occasion of the very beginnings of my own life. My father was born at the turn of the century, and the life of his father carries us back well into the nineteenth century. I am not sure whether this explained the long time-span of some narratives in my family. Maybe it was just the slow tick of the clock in the countryside, where few memorable events occur. About a certain valued possession, my parents would say, almost as first-hand knowledge, that it had been hidden in the cornfields, 'against the Cossacks'. I later found out that this must have happened in Napoleon's time. Instead of revealing the depth of time, such stories made me feel exposed to the endless plains and their horsemen.

These stories contributed to my earliest 'geopolitical vision', a cognitive achievement that does not necessarily require formal schooling or political awareness, as the real fathoming of history does. My vision certainly did not reflect post war geopolitical reality but probably contained more nineteenth-century reality than one might expect. A small town in the land of my family's tales was well-known for its regular (nineteenth-century) trade with Russia. This reminds us of the fact that feelings of 'nearness' are not a simple function of technical progress in transport and telecommunications. Russia in the nineteenth century may well have been closer to the world of country people in the eastern part of the Netherlands than later (Cold War) conditions would allow.

Stories about the world may differ between individuals, families, regions and nations. A survey of such differences at the micro-level would provide exciting material. In this book, however, more of a broad-brush approach is followed. This is not merely because the lack of survey data on the individual level necessitates such an approach. In spite of all the diverging personal experiences and family (hi)stories, certain events and pieces of information affect large groups of people in more or less similar ways. The nation state acts as an information system linking traumatic or joyous events in history to a particular territory. Identifying with a territory simply elicits certain views on the world, albeit in a contingent way, given certain national challenges, historical facts and ideals.

This book is an attempt to establish the occurrence of such views and to relate them to the history and the geographical conditions of nine nationalities or states. It is both an inventory and an experiment. As an inventory, I have to acknowledge that there are specialists, on each of the selected countries, who know more of its particular history or literature than I do. However, those specialists seldom tend to collect information on geopolitical visions and even when they do few of them feel called upon to connect this information with similar facts from other countries. That is why I have no hesitation in presenting this work, in spite of all its limitations. Fortunately, there are always studies of the foreign policy and national identity of particular countries which offer fragmented, sometimes unintentional glimpses of our theme. This book makes no claims to comprehensiveness, but if it results in further similar efforts, my aim will have been largely achieved.

As an experiment, I am ready to put this way of looking at nations to the test of the foreign policy decisions and international developments of the future. It is up to the reader to judge the value of this perspective for an understanding of coming events in these countries.

The beginning and the completion of an intellectual project depend on many circumstances over which an author has little control. I can easily imagine that this book would never have been written without my contacts with a number of people. I am grateful to Piet Schat for initiating me into the fascinating world of the German *Zeitschrift für Geopolitik* more than twenty-five years ago. Herman van der Wusten has always been stimulating with his continuous interest and great knowledge. I am further indebted to the following for their encouragement and advice: Martijn Roessingh, the editors of the Dutch journal *Internationale Spectator*, G.H. de Vos van Steenwijk (Moscow) and Ali H. Al-Sammak.

Finally, I have to acknowledge the invaluable help of my wife, Eva van Kempen, who has critically read through the whole manuscript and suggested countless improvements.

Gertjan Dijkink
Amsterdam, February 1996

THE NATIONAL EXPERIENCE OF PLACE

THE GENIUS OF PLACE

'Remember the words of Napoleon: Each state follows the politics of its geography.' These words of the French president François Mitterrand were meant to reassure his bewildered nationals after the sudden crumbling of the Berlin Wall in 1989.[1] The message that there is no news under the sun, and that each place on earth embodies a political destiny or mission, may be consoling under certain circumstances, but it evokes dark memories as well, not least for the French people. The Nazi predilection for a geopolitics (*Geopolitik*) presenting Germany as the natural ruler over continental Europe reminds us of the various and outrageous ways of translating place into politics. This episode in the history of thought dealt a fatal blow to all academic speculation on the political destiny of places. Yet for statesmen and nationalist thinkers alike, the theme has always kept the attraction of forbidden fruit. Half a century after the Second World War Mitterrand displayed no fear of becoming tainted with unsavoury ideas. But even his invocation of Napoleon cannot undo the extreme ambiguity of place as a political guide, and one might also doubt whether such a reflection on the course of European history since Napoleon is so reassuring for the French. It is a pity that the French President did not give any clues as to the contents of the politics he was reflecting on.

The dominant reaction to the troubled history of geopolitics has been to push geopolitical theories aside as nationalistic visions, myths or simply expressions of the capricious human will. But visions and myths are interesting in their own right. Since they reveal the human motivation behind some of the most abhorrent events the world has seen, they cannot be ignored. And since they draw their inspiration from geographical and historical facts, a more general perspective on their origins is worth trying out. The resurgent nationalisms and regionalisms of recent decades suggest that to live within a territory arouses particular but shared visions (narratives) of the meaning of one's place in the world and the global system.

Even if those who practice statecraft introduce fresh concepts about the outside world, these often rely for support on images and feelings which are deeply anchored in the life and experience of a particular group. 'Archetypes' and symbols which are regularly activated by politicians and the media have sunk to a semi- or subconscious level only through repetition and tradition. The prominence of frontiers and lines in American mental maps of the world, the idea of American purity,[2] or the gross distinction between East and West as opposite cultures,[3] are or were effective organizing elements because they never enter public discussion. They became, in the terms of discourse analysis, 'naturalized', part of common sense.

It has already been established that there are indications of 'political geographical socialization' in very young children. Educational researchers have subjected children (boys) aged from 7 to 17 in two countries, Britain and Argentina, to a territorial and national knowledge test.[4] They found that the English boys gave much more importance to the people and the principles that hold their nation together than they did to their national territory. The Argentinian boys, from as early as age 7, were much more prepared than their English contemporaries to look at national geography in political terms. It would be mere speculation to explain this difference by referring on the one hand to the history of the British Empire and on the other to the recent territorial struggle of Argentina, but it would not be a nonsensical hypothesis. The authors wisely refuse to enter into such speculation, although their view that it could probably be just as easy to find an explanation for the difference between the two nations if Britain had been a different country (for example one without a colonial history) is somewhat superficial. We can do without definite explanations because the systematic difference between the two national groups is already revealing enough. Obviously there are, at an early stage of life; cognitive features which may either be based on the particular history of a country and its environment, or influence how external events are judged. The test did not address the children's knowledge of other countries, but it is tenable that people who think about their nation as characterized by a set of principles will react to external events in a different way from people who identify with a territory in the first place.

Living somewhere means being exposed to the continuous stream of discourse produced by a local society *and* experiencing events which differ in kind from those happening elsewhere in the world. Both systems of information reinforce each other. The endless cycle of seasons watched by a subsistence farmer inspires a different view of the world from that of the town-dweller surrounded by projects competing for novelty. Each place suggests models for the world and provides blind spots as well. The prolonged absence of environmental threats may induce a false assessment of safety (both for one's own place and for the world as a whole), as has been demonstrated in the case of flood-plain dwellers. Kates coined the term 'prison of experience' in the context of hazard studies to describe this impact of the physical environment.[5]

In the present age with its possibilities of receiving information from anywhere on the globe, its cheap travel opportunities and homogenizing tendencies, to call our territorial context a 'prison of experience' seems an irresponsible overstatement. Yet there is still a basis for the thesis that persons born in this world are conditioned not only by the genes of their parents but also by the genius of a time and a place. There are at least three arguments to support this view. First, the culture-bound experiences of a person's formative years are difficult to extinguish in later life. Second, no amount of information can overcome the particular structure of information-processing pertaining to each place. The nature of press agencies, presence of libraries, journalistic fashions and training, are always present in information as a factor determining the nature of images. The idea that the emergence of an 'Internet' culture could sidestep such constraints is an illusion because we usually want and need local mediation of information. Third, the daily impact of events and human activities creates what psychologists have called an 'adaptation level', an unconscious standard for style, sounds, quantities and solutions. The clumsy interventions of the European powers in the first stage of the civil war in Yugoslavia were possibly affected by their inability to imagine that the violent conquest of territories (including 'ethnic cleansing') was from the beginning a real option for the parties to the conflict. This was completely at variance with the vision and experience of conflict that had developed in the European community in the post war era.

How experience and discourse together create an 'imaginative geography' of the outside world is a complex and fascinating story. Apart from a definition of the nature of geopolitical visions, this story requires some reflection on the territorial base which provides the vantage-point for constructing visions of the world: what kind of place, which social scale or political territory influences our images most profoundly? But first, I will make some comments on the tradition of geopolitics as a serious subject that has been taught in universities and military academies.

A SHORT NOTE ON CLASSICAL AND MODERN GEOPOLITICS

Geopolitics has been traditionally understood as the (scientific) assessment of geographic conditions underlying either the power (security) of a particular state or the balance of power in the global configuration of continents and oceans. Founders of classical geopolitics like Friedrich Ratzel (1844–1904), Alfred Mahan (1840–1914), Rudolf Kjellén (1864–1922) and Halford Mackinder (1861–1947) emphasized the natural advantages of certain locations in terms of land and sea power, or the 'biological' necessities in the spatial form and growth of states.

It is the latter tradition that fell into disrepute through the mutual understanding between German researchers in the fields of history, geography and planning and Nazi politicians during the 1930s. 'Mutual understanding' is the strongest wording that can be used, because Hitler's policies did not conform to the geopolitical conceptions (of Karl Haushofer and others) of the time. But even if Hitler had followed these contemporary theories and prescriptions, one might justifiably conclude that a disreputable history does not eliminate the sense of a 'geographic' assessment of national security. Indeed, such approaches continued after the war, except that the term 'geopolitics' was carefully avoided for one or two decades.[6]

There was another reason for the fading of the geopolitical perspective: the strategy of nuclear deterrence which developed since the start of the Cold War.[7] This shifted attention from territorial defence to the question of the nuclear *balance*. Captivated by questions about the potential for nuclear retaliation, few strategic thinkers were interested in spatial 'details'. A logical conclusion from this state of affairs would be to expect the rise of a new interest in geopolitics from the moment the bi-polar world order began to break up. Such a development actually occurred after 1970, and its corresponding vision became embodied in the writings and politics of Henry Kissinger.

Kissinger (born in pre-war Germany) stated that American foreign policy was informed by idealistic, pragmatic and legalistic traditions but no geopolitical tradition. The rise of Third World nationalism and the Sino-Soviet split had helped observers to see the endless nuances that could be introduced into foreign policy and the possibilities of creating a *modus vivendi* with regimes or ideologies which did not necessarily converge with the American ideal. In this sense, Kissinger's geopolitics resembled Bismarckian *Realpolitik*: the ideal of a steersman who guides his ship across the seas of international relations, who tries to gain by skilful anticipation of currents and waves, but who never falls victim to the illusion that he steers the waves themselves.

In contrast to classical geopolitics, Kissinger's ideas hardly referred to the kind of geographic principles associated with classical geopolitics. Implicit facts such as distance, territorial extension and demography were certainly appreciated, but the attraction of this new 'geopolitics' lay in its claim to picture the world as it really is, in its almost malicious delight in revealing the irrelevance of pious wishes and human perceptions. The geographic dimension was not absent, but usually it was a metaphorical evocation of all those international conditions which have to be acknowledged as 'resistant' rather than as a spatial theory. It is this sense of the term which gave it some popular appeal in titles such as 'the geopolitics of famine', 'the geopolitics of information', and so on.

The decline of geopolitics into disrepute as an academic subject after the Second World War, and the revival of a geopolitics with a very reduced geographical meaning à la Kissinger, inspired political geographers and other students of foreign policy to wonder whether some kind of geopolitics did not persist in foreign policy beliefs and political language, even if classical theories of land and sea power had become obsolete, through either political misusage or advanced weapon and transport technologies. Yves Lacoste and his French school of political geography, with their journal *Hérodote* (1976–), deserve credit for being the first to recognize the challenge of this new type of (critical) geopolitics. The French tradition of frankly acknowledging the roles of national interest and power, its 'j'accuse' tradition of intellectual criticism, and Lacoste's interest in the regional geography of South-east Asia, may all have played a part. In 1972 Lacoste suddenly achieved fame with his contention that the spatial pattern of the Vietnam bombings by the United States revealed the 'hidden strategy' of destroying the dikes in the Mekong Delta. Contradicting this view, official US statements identified the targets not as dikes but as military equipment and transports. In 1976 Lacoste published *La géographie, ça sert d'abord à faire la guerre*[8] ('Geography's First Use is for Waging War') in which he again denounced the public naivety with respect to the hidden (military) meaning of geographical data.

The American counterpart set off ten years later. Referring to American involvement in Central America, Ó Tuathail and others noticed the persistence of a language of geopolitics in foreign policy making.[9] Geopolitical language can be recognized by the occurrence of words referring to boundaries and the conflict between territorially bounded interests. The zero-sum view of international relations is more or less inherent in these approaches to the world. Another feature of geopolitical images mentioned by Ó Tuathail is the failure to appreciate the cultural or symbolic character of foreign policy aims. Whereas the French school began its revived study of geopolitics by stressing the disguised application of geographical knowledge in coercing people, the Americans began by pointing out that the relation between geopolitical language and real security or economic interests is sometimes rather dubious. Both American and French revisionist geopolitics owe much to the Vietnam War. Is this rudimentary geopolitical discourse about the world then only a fig leaf? No, says Ó Tuathail, it has to be understood in a much broader, more cultural sense. 'American foreign policy aims to perpetuate, secure and reaffirm the American way of life.' This may on the one hand involve pursuing hard economic objectives and on the other produce a foreign policy which simply 'enacts' the domestic ideals of identity. The latter approach constitutes the central perspective in post-structuralist studies such as Donald Campbell's *Writing Security*.[10]

Drawing much inspiration from the Cold War and the communist witch-hunt in the United States, Campbell observes that nation states feel a continuous need to re-establish their identity. Nation states should not be considered as entities with a fixed identity and corresponding external behaviour (foreign policy). We have to acknowledge the reverse too. Because nation states encompass ambiguous portions of the Earth's space and population, they always lack sufficient substance and consequently need foreign policy to define what is 'us' and 'them'. Foreign policy in this view is a 'boundary-producing' phenomenon rather than the outcome of a well-defined state. The antagonism of the Cold War aimed at disciplining American society (threatened by social-economic tensions) rather than defending the state against a territorial threat. The emphasis in certain countries on the drug trade as an external danger can be seen in the same light. Leaving aside the obvious criticism that this perspective will never be able to elucidate a country's entire foreign policy, we can accept two assumptions which have proven their value for understanding international events and national policies: national identity is continuously rewritten on the basis of external events; and foreign politics does not mechanically respond to real threats but to constructed dangers.

With these perspectives, the original aim of geopolitics, as a scientific explanation of the power and security policy of states on the basis of geographical position and/or natural resources, disappears from view. What remains is a security perspective of people who draw from their environment, physical and cultural, to construct visions of order and threat. This is not simply an elaboration of the traditional geopolitical approach through another variable (perception), but rather a theoretical adjustment to the new reality of the increasing permeability of borders and interdependency of states. In this new world with transnational economies and wandering minorities, the threats to peace and stability are coming from within rather than from without. External threats and internal doubts are hard to disentangle.[11] Even so, personal experiences of the world will be affected by the idea of a territorial self and its relationship with other territorial identities. Even in a world which does not simply correspond with the principles of a zero-sum game one may be feverishly searching for the game with the most satisfactory outcome.

LEVELS OF IDENTITY AND INTERPRETATION: CITY TO CONTINENT

The French writer Michel Butor observed in the 1960s:

> In my city many other cities are present, through many kinds of mediation: signposts, geography books, objects that arrive, newspapers writing about the world outside, images, movies showing me something about it, the memories I have, the novels which take me there.
>
> *The presence of the rest of the world has a particular structure in each place*, the relations of effective proximity may be quite different from the real nearness. I may only get informed about an event which happened a few meters away by means of a press agency, an editor and a printer 100 kilometres away.[12]

Butor talks about cities because they most closely approximate to his ideal of the world as a book. Removing the distinction between novel and world was one of the profound desires of the generation of French writers who created the 'modern novel' (*le nouveau roman*), of whom Butor is such an outstanding representative. They were excited by the structure of references and rhythm of activities characterizing a place and representing both a perspective on the world and a novel. Since the task of the writer is to investigate perspectives, a change of residence becomes almost a literary challenge. James Joyce's *Ulysses* (a city, a world, a book) is the great model in this tradition and to Butor and his colleagues the fact that it was written in exile is more than a casual fact.

The absence of allusions to the role of nationality in what Butor calls 'the presence of the rest of the world' is remarkable. It is undoubtedly characteristic of post war modernism, which has dispensed with its nineteenth-century bond with nationalism. In fact, the changes of residence Butor experienced himself as meaningful movements in his life largely involved a change in national context: he went to live and write in Scotland and Egypt for a while. But is it not obvious that the nation, as an information- and interpretation-system, is much more penetrating in its impact than a city or a local community? Butor's remark on the long detour in the stream of information identifies a feature of not only the modern technical world but also national information systems. Even more perplexing than what Butor himself observes, is the fact that sometimes we do *not* get information about events in the vicinity for the simple reason that they happen the other side of a national border.

Another, more deliberate, negation of the national context occurs in certain 'post-modern' discussions celebrating the new power of individuals or local communities to transcend the level of national politics. Chadwick Alger refers tirelessly to the Greenham Common protest of English women against nuclear army bases to illustrate his idea of (what I would call) unmediated trans-national commitment.[13] He is apparently unaware that the same example can be, and has been, used to illustrate the very opposite. William Bloom, in his study of the national identity dynamic, compared the success of the Greenham Common women with the utter failure of the movement for British unilateral disarmament.[14] How can we explain this glaring contrast between success and failure in the two movements? According to Bloom, it is precisely the national identity dynamic, the possibility of translating an issue into Us-and-Them terms, which is the prime factor in the Greenham Common success. The US bases could be represented more or less as a foreign occupation of national territory, whereas the disarmament movement had to resort to the language of universalism ('world peace'). The Greenham movement might not have received the immediate support that it did if it had been focused against *British* nuclear establishments!

Us-and-Them opposition on a very wide scale is discussed by Edward Said in his study of Orientalism. Said argues that centuries of fear and domination of the 'Orient' by European powers have left an almost indelible mark on the sphere of knowledge that became known as 'Orientalism'.[15] This knowledge, incorporated in texts produced by scholars, politicians, fiction writers and travellers, almost never entered into a real dialogue with the human subjects concerned. It stressed the unsurmountable difference between the 'East' and the 'West', particularly in terms of cultural personality and social structure. The East was described as sensual and inclined to despotism; Arabs were devoid of energy and initiative, inaccurate and deficient in the logical faculties, and so on. This field of learning and observations, 'because generated out of strength, represented and created the Orient.' In this view Said joins Foucault, who likewise stressed the knowledge-producing features of systems of domination. The main thrust of Said's argument is that even the kind of interest that purported to be driven by the selfless curiosity of scholars, or by admiration for other cultures, ultimately produced political ways of looking at things.

Said's book has been deservedly praised but it also elicited (and still elicits) hostile reactions. Some of these are understandable, although they unjustly reproach the author for saying or omitting something which he never aimed to say or do. The resentment is caused by the fact that, as many readers probably feel, the approach seems to narrow the scope for inter-cultural criticism. The message of *Orientalism* was not very well attuned to the changing spirit of the times. It appeared precisely at the moment when many Western intellectuals were beginning to reproach themselves over having accepted political changes in China (the Cultural Revolution and Mao's Great Leap Forward) as something not susceptible to Western criticism (manipulation). After the

Chinese themselves became increasingly critical of this turn in their history, these intellectuals discovered that human judgements across cultural borders are possible. The same criticism is levelled at Said by Kanan Makiya, who believes that the idea of Orientalism has provided Arab intellectuals with an alibi for keeping silent about the violation of human rights in the Arab world (particularly in Iraq).[16] As Said rightly answers, his aim was never to pass moral judgements on either culture.[17] He simply wanted to analyse a system of knowledge production.

Yet there is a philosophical question behind this controversy which gets more to the heart of the matter. The message of 'Orientalism' is that all generalization concerning the Orient is biased because it delimits a geographical area which is only a unit (a virtual counter-world) in terms of the political relations with imperialistic powers. Is it then allowed to delimit another world (Europe) and identify it with a systematic field of knowledge about the East? And is it not the case that much of the discourse identified by Said as 'Orientalist' belong to the type of knowledge (or rather 'gossip') which is generated in every place where people encounter an external world considered inferior?

'Orientalist' visions have developed in various countries with and towards neighbours with a different culture or living standard. At the turn of the century the Germans considered the world beyond their eastern border (officially Russia, but actually the Polish and Russian nations) to be one with rough and still unappeased manners and customs (F. Ratzel, see Chapter 2, Note 1). The Poles were depicted as unable to create or conduct a rational economy. The Poles, on the other hand, considered themselves (because they shared in the Roman Catholic heritage) as the last outpost of an age-old European culture. The Russians themselves did likewise, with their missionary attitude towards the Giliaks of the lower Amur River and other people in the Far East.[18]

The picture becomes still more confusing when we know that Russian intellectuals in the nineteenth century and German ideologues in the first decades of this century characterized their own spiritual world as 'Oriental' (or Asian) in comparison with the French or Anglo-Saxon mentality. The imperialist role of Britain and France, and the German frustration on reaching statehood at the time when the world seemed largely divided between those two powers, was of course the main motive behind these ideas in Germany, not the real cultural gap between those nations. But the self-analysis of Russian and German intellectuals in some respects painstakingly established national qualities identified in 'Orientalism' as negative (European) qualifications. The West European mind was called 'rationalistic' and 'materialistic', whereas the German (or Central European) mind was praised as 'based on insight', even as 'mystic'. In the light of such facts the restriction of Said's research to (mainly) British and French sources has more than a practical significance.

I do not claim that the level of identification of those who produced 'Orientalist' knowledge always and exclusively corresponded with the nation state. The examples given above indicate that a kind of European identification may have been implicated in the feeling of belonging to the Christian community of faith. And the same may apply to membership of the community of scientists, or people working in a technical profession. But even in these realms the idea of a special national mission or responsibility was often present. Moreover, imperialism, the system underlying Orientalism according to Said, was never a supranational enterprise. Every European royal family may have sent its representatives to the opening ceremonies of the Suez Canal (1869), but fine words about 'Europe's connection with the East' could not alter the fact that to the last the British had obstructed the Suez enterprise.

Nor does the fact that there are important similarities in the world views of different European countries, justify the choice of a European level of analysis in studying geopolitical visions. One might expect a European geopolitical view to emerge only if all or a number of European countries encountered external forces or dangers. The Second World War and the 'occupation' of Europe by foreign forces (the United States and the Soviet Union) is the closest to such an experience in modern history. Hervé Varenne, reflecting on the possibility of a European nationalism, observes, 'It may not be insignificant that the first institutionalization of Europe coincided with the first military occupation of Europe by non-Europeans.'.[19] Future historians may well conclude that this event marked the start of a process in which feelings of kinship between the United States and European nations gradually gave way to more contrasting identities which would inevitably reinforce Europe as a geopolitical unity. Until now, however, the national perspectives of the different countries, although sharing many common (European) features, are still strong. Moreover, the unqualified support of Western European publics for the maintenance of friendly relations with the United States did not – until recently – betray a longing for a fundamental redefinition of geopolitical schemata.[20]

The situation was different in Eastern Europe, precisely because the intelligentsia in those countries felt themselves cut off from mainstream European culture – 'kidnapped', as the Czech writer Milan Kundera called it, in a celebrated essay on the predicament of Eastern Europe (1983).[21] He quotes the desperate voice of the director of the Hungarian radio station who, during the 1956 Rising, shouted to the world: 'We are going to die for Hungary and for Europe' (he actually was killed). Nowhere in the capitals of Western Europe, nor in Moscow or Leningrad, could anybody have thought up that phrase, opines Kundera. His essay confirms the necessity of an alien threat to evoke supra-national identification. One wonders whether the feelings of East Europeans have become more 'nationalized' like those of West Europeans in the years since 1989.

New dangers – whether real or imagined – which might reinforce conceptions of Europe as a unit are the phenomena of migration to Europe and Islamic fundamentalism. The increasing pressure of migration has led to some policy coordination between a number of EU countries (the Schengen Pact). In the American press these moves were depicted as creating a 'Fortress Europe'.[22] It is too early to establish if these events will have any fundamental influence on the outlook of both foreign policy elites and the general public in European countries. The same obtains for Islamic fundamentalism, which is hardly recognizable as a unified international player, and, moreover, has never made Europe a special target.

At the end of his essay on European nationalism, Varenne draws a devastating conclusion: nowhere does he feel the presence of Europe in the suburbs of Dublin in the way one can be aware of the United States in the suburbs of New York.[23] This sounds like a sour comment on Michel Butor's high-pitched eulogy on the city as mirror of the world. Butor would certainly have found *a* Europe in Dublin, the holy city of the modern novelist. But the messages are not really contradictory: there is another layer of references which drowns the European presence in Ireland or, for that matter, in every other European country. It in fact means that there may be an Irish Europe or even an Irish World but not a European World in Dublin.

Such observations seem in accordance with the results of public opinion polls or turn outs at elections for the European parliament, all of which show a minor role for the European as opposed to the national political scene. In a series of surveys spanning the period 1972–1988, people living in the Italian border region of Friuli were asked about their attachment to different spatial levels ranging from the neighbourhood to the world. The highest attachment was first to the local community and second to the nation state. Europe scored very low in five of the six studies.[24]

Yet there is another phenomenon which requires some comment in this context. Public opinion polls have also shown very low levels of national pride (the 'willingness to die for your country', etc.) in a number of European countries since 1970: in particular Germany, Belgium, the Netherlands, Luxembourg and even France. It is probably no accident that they are the oldest members and core countries of the EU. They also contain the zone which witnessed an early (fifteenth-century) development of small-scale political life centring on merchant-cities along the Rhineaxis.[25] Nowadays these areas again show a degree of regional awareness which at the same time aspires to internationalism in economic affairs.[26] Does this mean that national identity loses its meaning as an international frame of reference? It is not very likely that the inhabitants of Baden-Württemberg will construct their world map according to the structure and interests of a multinational firm such as Robert Bosch. As long as the schools in Baden-Württemberg teach German history, and as long as they are reached by a national press and TV channels, the interpretation of the world map will be predominantly national.[27] The low level of nationalist feeling primarily tells us something about the well-defined and comparatively stable political order after the Second World War in this part of Europe, and about the satisfactory course of economic affairs. They have undoubtedly fostered the idea of European integration with the region as the basic administrative unit, but this does not confer the cultural and political meaning on both levels which necessarily underlies any geopolitical vision.

I have dwelled at such length on European countries that I have probably laid myself open to the charge of 'Eurocentrism'. But there is good reason to start with Europe in discussing the future of national identity. For it is the part of the world where the nation state originated, developed and now seems to be withering. There is little agreement on the new forms it will take in the future but that a basic change is taking place in the existential world of the Europeans seems beyond doubt. Everybody who compares long-term trends in attitude/opinion surveys in Europe and the United States will be struck by the fluctuations in the former and the comparatively stable scores in the latter.[28] Whether this is a new phenomenon or not is difficult to tell. But it does suggest the 'turbulence' of systems which cannot rely on their own powers to remain unaffected by environmental change.

THE NATURE OF GEOPOLITICAL VISIONS

In order to maintain a broad view and to keep my argument uninterrupted, I have left the concept of geopolitical vision undefined as something intuitively clear, something like 'imaginative geography'. But now we have to face its precise nature. What kind of cognitive act or discourse is covered by this concept? How much can it be separated from the discussion of national identity in general? How does it differ from the 'policy belief system' studied by foreign policy analysts?

Choosing a definition is far and away the easiest part of the task. I define geopolitical vision as: *any idea concerning the relation between one's own and other places, involving feelings of (in)security or (dis)advantage (and/or) invoking ideas about a collective mission or foreign policy strategy.* Notice that I do not deem it necessary to introduce the concept of the state (although 'foreign policy' more or less implies it) or to stipulate that the vision be shared by the majority of a national population. The state can sometimes be an 'external' source of insecurity feelings for those who live within its territorial borders and there may be alternative geopolitical visions within a nation state. To be more explicit: I would not exclude the Kurdish people from having distinct geopolitical visions. A geopolitical vision requires at least a Them-and-Us distinction and emotional attachment to a place.[29]

Whether an analysis of geopolitical visions can be separated from a thorough treatment of national identity is a more difficult question. Anthony Smith has listed several features of national identity: an historic territory, common myths and historic memories, a mass culture, a common economy and common legal rights and duties for all members.[30] All this 'commonality' inevitably supposes discontinuity with other groups, even if they are racially and culturally very similar to the nation concerned. The 'historic territory' implies a narrative of conquest, defence, liberation and loss in which again certain 'Others' play a role. National identity can hardly be imagined without the feelings of trauma and pride that arise from external relations. In this respect feelings of national identity and geopolitical visions are difficult to separate. Nevertheless, geopolitical visions are more the concrete translations of such feelings into models of the world.

All discourse that emphasizes the unique British traditions of freedom, justice and pragmatism, in contrast to 'Continental' corruption and rhetoric, may be considered as material for building the British national identity. But to present an ideal of the British position in Europe by referring to the position of the Crown Colony of Hong Kong *vis-à-vis* China, as Thatcher's Environment Minister Nicholas Ridley did in 1990, is to state a geopolitical vision in the narrow sense.[31] It was neither an appropriate analogy nor a wise idea, but as an analogy it appealed to the national-identity feature of freedom. By connecting it with the preference of firms for a low level of national economic control, an element of national advantage (security) is introduced which turns it into a genuine geopolitical vision.

To outline the 'ideal type' of a geopolitical vision requires outlining the structure of an account of the nation in its relations with the external world. This would imply, first, some justification of the 'naturalness' of the *territorial borders*. It could involve an exaggerated account of homogeneity within the borders compared with dissimilarities beyond them. It might also stress the natural unity and its human adaptations in (pre-)historic times, or the natural character of the borders, including those which have become 'natural' through the shedding of blood in their defence. Another approach would be to highlight a *core area* (usually including the capital) as a physical and spiritual force which continually reinforces national unity.

Then there is a *geopolitical code* described by John Gaddis in his study of American security policy as 'assumptions about [American] interests in the world, potential threats to them, and feasible responses'.[32] Let us call it mundanely: a list of friendly and hostile nations. The American containment policy towards communism during the Cold War relied on such lists. However, each new administration tends to devise a new geopolitical code designating different countries as hostile or friendly.

A closely related phenomenon is the tendency to choose other countries (and their international relations) as a *model* to follow or to reject. Young, ambitious nations, such as Germany and Argentina in the nineteenth century, have gone through several fashions in copying the culture and institutions of other countries (for example, those of England, France, or the United States). But often the attitude was reversed after a time, when dependency on a foreign example was perceived as a proof of non-identity. Then the gaze turned inward, searching in history for primitive roots and a pure local identity.

A fourth device is the idea of a *national mission* which may entail imperialist aims, playing the role of gatekeeper for a hinterland, preserving a historic heritage with a universal significance (like the Muslims' holy of holies, the Kaaba in Mecca), educating a revolutionary class, and so on. Here the model function may get reversed, in that the nation considers itself as the standard for what happens elsewhere. Exciting events in national history, war and revolution, tend to loom so large in the interpretation of the world that any awareness of the possible influence of other social or physical conditions is blotted out. The Cuban revolution, initiated by a handful of guerillas hiding in the Sierra Maestra, must have been a first-rate 'peak-experience', almost a miracle (Figure 1). However, had it really remained a miracle in the national perception, then some policy failures might have been averted. One of these failures, leading to the death of the hero, was Che Guevara's mission in Bolivia. Were he and his fellow revolutionaries really so naive as to believe that a small band of guerillas hiding in the Andes mountains could unchain a

Géographmes illustrés: la Plaine, la Ville et la Sierra...

Figure 1 How the French journal *Hérodote* pictures the geographical visions of the Cuban revolutionaries (Jean Hervé 1977 in *Hérodote* vol. 5. Reproduced with permission by *Hérodote* – Éd. La Decouverte)

revolution like Cuba's? This does seem to be so, because nowhere do Guevara's diaries account for the different social conditions and the completely different attitude of the peasants in Bolivia.[33]

Finally, one might include in the category of geopolitical visions assumptions about *impersonal (even Divine) forces*. This may include laws of change and forces organizing the world that can be taken advantage of or that require containment, such as modernization, Islamic fundamentalism, free enterprise (the world of multinational corporations), splendid isolation, and so on. A large part of classical geopolitics consists of statements fitting in this category. The balance between sea power and the power contained in the Heartland (Mackinder), the idea of the mysterious unity of *Mitteleuropa*, and the domino theory of communist expansion, were such visions.

Having uncovered the relationship between national identity and geopolitical visions – by understanding geopolitical visions as translations of national-identity concepts in geographical terms and symbols – we now come to another closely related concept: the foreign policy belief-system. Described as an 'all-encompassing set of lenses through which [decision-makers] perceive their environment,'[34] the foreign policy belief-system may easily cover most of the subjects introduced in the preceding discussion. In the study of international relations 'belief-system theory' is often presented as either an alternative or in addition to other explanations of foreign policy decision-making, such as theories of coalition-building or leadership competition in domestic politics.

Douglas Blum has given an illuminating example by comparing the results of such analyses of Soviet foreign politics with the insights that belief-system theory can provide.[35] The belief-system approach emerges from his discussion as a necessary contribution to an understanding of why Soviet foreign policy remained so unchanging, in spite of negative outcomes and significant changes in leadership after Stalin and Khrushchev. The intrinsic durability of beliefs, particularly '*core*-beliefs', is put forward as the key to this failure to change. Once adjustment does become unavoidable, it initially tends to change only conceptions of the world of lesser scope ('*intermediate* beliefs'). Only the persistence of problems over a longer period will change core-beliefs. This is what happened under Gorbachev.

One of the elements in this mental scheme is the self-image of the Soviet state-elite. The core-belief under Stalin presented the Communist Party of the Soviet Union as infallible and as the sole arbiter of political legitimacy within the communist world. The lower level translation of this (an intermediate belief) was a narcissistic belief in the universal relevance of Soviet experience. After Stalin's death this intermediate belief was changed in the sense that 'separate roads to socialism' were acknowledged. But this did not imply that the (core) belief – the infallible judgement on right or wrong – was abandoned. The experiments in Hungary (1956) and Czechoslovakia (1968) bear witness to a continued will to interfere with policies that had been prudently announced as 'separate roads to communism'. Only under Gorbachev did the self-criticism emerge that frankly admitted the occurrence of foreign policy failures in the past.

Geopolitical visions may be described as just a subset within the foreign policy belief-system. They may consist of volatile conceptions of international relations or be conceived as core-beliefs with even deeper roots then those dug out in the usual analysis of foreign policy beliefs, which often relies on well-articulated policy options and easily verbalized visions. It is remarkable that not only the core-belief quoted above, but all the other core-beliefs identified by Blum refer to more or less general communist convictions. This is not necessarily wrong. The 'all-encompassing' influence of peak-experiences such as war and revolution on patterns of thought is one of the assumptions of this book. Yet this overwhelming presence of communism in the Soviet belief-system may also be imputed to the viewer experiencing the Soviet Union in the context of an antagonistic system and its recent demise. It would be interesting to speculate whether there are any Soviet or Russian assumptions about the world with roots that are deeper and more determined by place than is communist ideology. Such visions would possibly be disguised in communist clothing, or otherwise suppressed, albeit without losing their vitality.

'Even contemporary images of Asia applied by Russian intelligentsia contain little of the obligatory Marxist-Leninist dosage of ideology,' wrote Milan Hauner in the latter days of the Soviet empire.[36] Hauner asserts that the reality of the Eurasian landmass has always challenged Russian and Soviet politics in a way that could not be incorporated in Marxist ideology. Lenin's relocation of the Russian capital from St Petersburg to Moscow, the relocation of borders and spheres of influence after the Second World War, the Sino-Soviet split, and the invasion of Afghanistan, were all indications that Soviet policy after the revolution was preoccupied with the problems of maintaining the integrity of the empire, materially and ideologically. Despite the fact that the subject of 'geopolitics' was never debatable, geopolitical considerations, rather than the aims of world revolution, controlled external relations.

This discussion brings us back to the words of François Mitterrand and, through him, of Napoleon. It is undoubtedly too much to expect a mechanical link between national political visions and geography. Geography only acquires meaning through historical events and through a way of life that geographical determinists alone have tried to explain directly on the basis of the physical environment. Yet one may at least ask whether the development of geopolitical visions follows certain patterns, independently of political changes; which type of vision is most perishable; and what and whose needs are satisfied by this type of geographical imagination?

THE ARGUMENT IN BRIEF

Each human being yearns for a kind of world order, a sensible pattern of people, things and behaviour in the world. This disposition involves a defensive response to changes, not merely a response to any arbitrary change but particularly a sensitivity to events that conflict with one's vision of order. An example of such a response might be: 'Asian industry cannot compete with European quality.' The standards for what comprises order are acquired during one's youth through exposure to events and objects in the environment and through becoming a member of an interpretive (signifying) community. Threats to the perceived order arouse feelings of insecurity and severe disruption of the order may even produce personal and social pathologies.

Because visions of order are derived from a community and from experiencing an environment (be it city, country, wilderness or island), it is to be expected that such visions are shared by the members of a localized group. Since individuals usually belong to several, vertically and horizontally differentiated, communities and environments, difference and similarity will occur at the same time. Whether difference or similarity is relevant in a given situation will depend on the kind of challenge, the disruptive force. For example, a significant improvement in the relation between two countries may typically invoke nation-wide visions of threat in a third country. Nation-wide does not mean that the vision is shared by the entire population, only that it does not coincide with any other group difference.

The national idea still represents the most relevant recourse for those who wish to restore order when they are faced with a serious disturbance, whether the threat comes from without or within (the state). It has even been asserted that foreign policy is a way to redress domestic order rather than a reaction to objective changes in the international system. Visions of order that, explicitly or implicitly, appeal to outward reactions of a nation's institutions (for example, government, diplomacy), are called *geopolitical visions* in this book.

The ideal reaction to a disturbance is its elimination. Since this will often be an unattainable goal for the responsible institution (either because of a lack of the necessary power or simply because the disturbance only dwells in people's minds), the restoration of order will tend to be primarily symbolic. This means that the (changed) order is simply given a new meaning by reconstructing national identity. Such self-absorbed attitudes suggest that geopolitical visions have little to tell about real international happenings in this world. It would have been possible to deduce a completely wrong expectation of future conflicts by analysing the American mind between the two world wars. Public opinion was focused on the danger of Soviet communism rather than on countering Germany and Japan. But this insight would at least have allowed one conclusion concerning the future: that Americans were unprepared for certain wars (namely the next one). Global events like the dissolution of the Soviet empire have often taken even the most ingrained diplomats by surprise. Yet, reactions of countries and people to such events appear less whimsical if one accounts for geopolitical visions. Visions undoubtedly produce facts but we have to learn to deal with these facts in language other than the one 'spoken' by the visions themselves.

THE COUNTRY OF *ANGST*

(Germany)

The German–Russian border is not the border between two states, but between two worlds.

(Friedrich Ratzel, 1898)[1]

Of all the Europeans the Germans know more angst than other peoples.

(Helmut Schmidt, 1987)[2]

Immediately at Germany's eastern border a large earthquake zone begins, extending to Vladivostok.

(Jochen Thies, 1993)[3]

FEAR IN 'THE MIDDLE'

For few western states did the end of the Cold War imply so drastic a revolution in the geopolitical situation as for the Federal Republic of Germany. It is not surprising that the new notes that sound in the unifying Germany should fall on deaf ears elsewhere. In the United States particularly, a country which supported German unification but never experienced similar power shifts on its own borders,[4] irritation arose about Germany's infatuation with Eastern Europe and its lack of concern with problems outside Europe such as the crisis in the Gulf.

Outside Germany, a new need was felt for a debate about the continuities in German foreign policy, or, to put it more basically, for an understanding of Germany's existential situation within a long-term perspective, at least a perspective extending beyond the Cold War period. An alert American essay, addressing the American policy community, appeared in 1990. The author, Walter Russell Mead, advanced a crucial difference between Germany and its neighbours: 'it may be that German politics has been filled with more angst and questioning than Holland's precisely because Germany's immediate surroundings are so much more complex.'[5] Helmut Schmidt, the ex-chancellor, however attributed German angst in his memoirs (1987) to postwar events: the division of Germany, the presence of foreign troops on German soil, and the lack of freedom in East Germany. Mead does not quote sources to elucidate the background of his ideas on the 'complexity of the surroundings', yet the viewpoint is not new. It fits into a tradition of geographical explanations for Germany's political behaviour, particularly the school of ('right-wing')

German historians who tried to explain both world wars with reference to Germany's geopolitical situation.[6] The historian Wendt, for example, calls this angst an 'essential and continuous social psychological basic way of being in Germany', related to the central location of Germany.[7] Germany's geographical location in Europe would entail the risk of a war on two fronts. The idea of a preventive war, so provocative to other European nations, developed with this situation in mind.

One does not have to agree with these authors about the compulsive effect of geographical locations. It is worth quoting what the Sprouts established already long ago:

> it is fruitful to distinguish analytically between the relation of environmental factors to policy decisions on the one hand, and to the operational results of decisions on the other. With respect to policy-making and the content of policy decisions ... what matters is how the policy-maker imagines the milieu to be, and not how it actually is. With respect to the operational results of decisions, what matters is how things are, and not how the policy-maker imagines them to be.[8]

Mead is careful enough to add other ingredients to the German question: its great economic power, and the German experience of being an 'outsider' in international politics. The seeds for this feeling were sown by Germany's late emergence on the international stage, in 1871. In a situation already dominated by an established balance between other Great Powers, the idea could take root that these all begrudged Germany the right to exist. The first international reactions to Germany's unification, however, were certainly not dismissive.

If we had to suggest realistic causes for the German experience of being an outsider state, then two world wars and National Socialism would present sufficient arguments. However, the occurrence of these wars and the movements in their wake are usually explained by German feelings of isolation. Therefore, any more profound explanation of the German world view should go back at least to the social and political situation at the end of the nineteenth century, immediately before and after German unification. Have fear and uncertainty about the political geographical structure of Europe been ingredients of the national experience from the inception of the German state? And did such feelings also invoke aggressive plans?

We could expect no more direct answer to such questions than what is to be found in German treatises on political geography, a discipline which had just been invented in the new Germany. Friedrich Ratzel was its driving force, a geographer who became famous and notorious with his concept of *Lebensraum*, the natural need of (young) states for physical space. The political geographical conceptions of Ratzel were quite well attuned to the problems of the German state and its statesmen at the turn of the century. But for ideas which also fit in with public opinion we must refer to one of his real popular works, *Deutschland: Einführung in die Heimatkunde* (1898).

In our time, states Ratzel in the preface to his book, for many Germans there are hardly foreign countries in Europe any more. Therefore the German has to know what his own country means to him and particularly how land and people relate to each other. It is true (in 1898) that the Germans have ventured into areas far removed from the current state borders, to the east and the southeast, but all such colonization waves and conquests have been extinguished in the course of time. Thus the habitat of the Germans today (1898) is also the place which Tacitus determined as the German territory: between the Rhine and the Weichsel and (for the later Romans) between the Alps and the North Sea. Ratzel says: 'There is a comforting element in the thought – which the Germans should not forget(!) – that the life of this nation was most vital when it narrowly fitted this core area.'

This statement sounds anything but imperialistic and aggressive. But it also has ring of military vigilance and suspicion. According to Ratzel, one cannot give France, Italy and Spain much credit for having maintained their current size and geographical shape, but for the Germans it is a sign of extraordinary power and endurance. After all, the Germans are surrounded by neighbours, and their country lacks clear natural borders. Therefore, Germany for a long time constituted the no man's land where its neighbours settled their military disputes. For Ratzel Bismarck proved an excellent political geographer when he spoke these words in the Reichstag in 1888: 'God has placed us in a position in which we are prevented by our neighbours from slipping into dullness and inertia' (p. 18). Germany only exists because it is (militarily) powerful.

Germany's borders are a problem, but the eastern border is a special case in Ratzel's view. The process of the formation of states occurred in Europe from west to east. France finished its state development earlier than Germany because it encountered the natural borders of the Rhine and Alps. The *Ostwachstum* (expansion to the east) of Germany and Austria has taken place in Slavic territory, but is not yet accomplished. Germany remains open to the east. Besides, this border implies a 'cultural fall' (like the fall in a river). France is securely enclosed by sea and mountains and areas with a cognate European culture but Germany has one side which is exposed to the 'cold air current of a crude, not very mature, not yet mollified popular life . . . ' (p. 304).

Whereas in one section Ratzel characterizes the German expansion of the distant past as unorganic (not related to the German primeval earth), else-where his treatise appears to offer scope for developments which put the German house in order (particularly in the east). Whereas natural borders assume an almost deterministic significance for the accomplishment of a state (France) in one place, elsewhere they are subordinated to laws of a more cultural nature. Ratzel calls the border with the Netherlands the most 'un-organic' and 'worst' border of Germany, but because the country is politically neutral and the Dutch legitimately consider themselves as a 'separate people' (*Sondervolk*), one can conceive of territories such as the Netherlands as large lakes on the German border.[9] The eastern border is different. Here, the author seems rather to imagine the kind of border which is attributed to historical empires: zones of transition to the world of barbarians, a kind of atmosphere becoming thinner and thinner.

In spite of this position of Germany at the edge of the European culture area,

Ratzel shares the idea, advocated by contemporary historians like Heinrich von Treitschke, of the cosmopolitan character of German culture. According to this, as a latecomer on the international stage, the German nation had watched other nations intently and with admiration. It had adopted the best qualities of the French, the English and the Americans but, consequently, it had become unselfish and unsure about its own identity. In Germany the people had always kept their minds open to the literature and art of other nations ('mentally spanning the whole world'), but the drawback was a lack of national identity, coupled with an excess of local patriotism – all this again according to Ratzel.

Ratzel's book is an almost perfect reflection of the blend of expansive liberal thinking, economic self-satisfaction and distress rooted in the European political geography that characterized the mood of German society at the end of the nineteenth century. National-existential perspectives reverted repeatedly to the fact of Germany's central location (*Mittellage*), but with paradoxical outcomes: on the one hand Germany was seen as a spiritual marketplace, a crossroads of international movements where ideas are exchanged and new impulses originate; on the other hand, the pressure of the surrounding world was constantly felt, restricting the German *Lebensraum* in a way that had to be resisted. The Darwinist implications of this concept are absent from a book like *Deutschland* . . . , which addresses a broad public. Such meanings are to be found between the lines rather than in direct argumentation. Ratzel's description of domestic society and international relations is in an idiom that in later times would be called 'social physics'. This approach describes social and cultural facts as processes like those occurring in liquids or gases, with the corresponding vocabulary of pressure differences, current, and cooling, or as analagous to the hydrological phenomena of tides and the fall in rivers. This may have been a literary device, a way to stimulate the imagination that was bound to strike a chord in the century and the country that displayed such astonishing progress in physics. But Ratzel, like other writers of the time, does not always make clear how literally his rhetoric and metaphors should be taken. In this respect these writings are also located at a crossroads: the transition from the creative and explorative period of the *Gründerzeit* to the horrifying dogmas of National Socialism and the holocaust.

REACTION AND RANCOUR

What do Richard Wagner (1813–1883), Karl Marx (1818–1883) and Karl May (1842–1912) have in common? Apart from the fact that they were Germans and that their work is probably symptomatic for the second half of the nineteenth century, their worlds hardly seem to touch. Nevertheless, they constitute a unique group which is difficult to complete with others: all three were German creative geniuses of the nineteenth-century and each in his own field produced a great, perhaps the absolute, twentieth-century bestseller and box-office success.[10] Karl May may not be taken very seriously as a literary writer, but his commercial success places him on the same level as the others.[11]

Commercial success does not seem a very intelligent way of classifying intellectuals and artists. It takes something more to be able to subsume them in the same category. I have lumped these three figures together because they seem to represent the range of spiritual currents in German culture at the end of the nineteenth century. There is a further common feature: few creative geniuses in the world have succeeded at the same time in both gathering so many admirers and in antagonizing so many others. The ups and downs of Marxism require no comment. In his own time Richard Wagner's music evoked already embittered controversies between Wagnerians and Brahms-supporters in Germany and France alike. Later Wagner's anti-semitism and the admiration for his work in National Socialist circles was to lead to a taboo on the performance of his music in Israel. Karl May's adventure stories lay on Hitler's bedside table, one accolade that reinforced the disgust the writer already evoked in literary circles by his reliance on the faculties of supermen.[12] Together, these three creative personalities seem to incarnate A.J.P. Taylor's view that in Germany, land of the middle, in fact things could not be kept to the middle of the road; everything seemed to incline to extremism.[13] The seeds of the controversies that their works evoked in later times had been possibly sown by impulses that already overshadowed their own lives. All three rebelled against society, a resistance ranging from criticism to rancour. Marx and Wagner had been carried away by the revolution of 1848 and they suffered, each in his own way, from the trauma of revolutionary failure and the return of authoritarian society. Both of them had to flee the country to avoid arrest and prosecution. In the new Germany, Wagner was hurt again by the Reich's meagre support for his musical theatre and Bayreuth festival plans. For a second time exile seemed the only possible response; this time the refuge would be North America whose 'classless' society, he thought, would welcome his work more warmly.

It is not surprising that Karl May's Saxon hero, Old Shatterhand, should once, in a hallucination, have heard the 'future music of Richard Wagner' on the American prairie. That was a mistake; nor was May any supporter of revolution. He had his central figure, Winnetou, educated by Klekih-petra (white father), a mysterious European who later told Old Shatterhand in confidence that he had been one of the agitators in the Revolution of 1848, someone who administered the 'drug' that robbed the masses of their senses. Klekih-petra had become repentant and made the self-diagnosis that 'to the man who doesn't know God, neither King nor government is sacred.' In this way, even the savage Indian owed his – Christian and bourgeois – morality to 1848. 'Since we read your works,' some readers wrote to May, 'we aren't social-democrats any more.'[14]

Karl May gained his own traumatic experience at the hands of the law. He was sentenced to imprisonment for petty theft; there was no political background. It is not surprising that material property seldom brings about good in May's writings. In his novels treasures, gold and silver frequently crop up. They are either lost, or bring down a curse that finally crushes the owner. The author's anti-materialism clearly has a personal background, but it also concurs wonderfully with the spirit of the period after 1871, when many people were searching for a new spiritual certainty amid the disruption and uncertainty caused by abruptly changing personal fortunes. Members of the lower middle class were annoyed by the *nouveaux riches*, or had themselves fallen victim to the stock market crash of 1873. For these common citizens, the rejection of wealth, and the ideal of an ordinary compatriot who first, as greenhorn, becomes a laughing-stock but later reveals himself as Superman, was a consolation.[15]

This image also applied in the international setting. Germany was, of course, a new arrival on the international stage, with a head full of armchair learning, and never having held a gun. Germany would teach the other European countries a lesson in unselfishness, but also one in power and intuition. May's heroes are always Germans (but more precisely, Saxons!), whereas the unscrupulous scoundrels, driven by gold fever, are pre-eminently Anglo-Saxon, sometimes French, adventurers. The delineation of the differences in national behaviour even extends to philosophy. Anglo-Saxons make use of blunt empiricism (''t is clear' is the stopgap of one Anglo-Saxon character), whereas the German hero Old Shatterhand proves his genius always by introspection (*Einfühlung*), in his ability to see what cannot be seen but turns out to be highly relevant.[16]

The negative impulse in Germany's search for identity, denying other ideas, opposing other classes, nations or preceding generations, has been attributed to the lack of durable political institutions.[17] During the nineteenth and for part of the twentieth century, Germany was a political territory without stable institutions such as a democratic parliament, national council or other 'public' authorities. In the nineteenth century Germany watched other countries in order to construct its own national identity, but each generation tended to reject the intellectual constructs of the one before it. As a consequence of the lack of institutions to absorb political energy, to visualize society and its working, self-definition remained a merely intellectual act.

Thus, for the Germans national identification started in the early nineteenth century with the cultural ideal of a small, civilized middle class. The aim was to form purveyors of culture (*Bildung*), citizens with an intimate knowledge of philosophy and classic craftsmanship in poetry, architecture and painting. However, succeeding generations attempted to extend this elitist goal to interests with a broader appeal, such as the *Turnvereine* (gymnastics clubs). The concept of *völkisch* (to refer to some kind of naturalness and purity) became a measure of the quality of the national ideals pursued. The failed revolutions of 1848 brought everyone down to earth. Cultural and political ideals proved to lead nowhere, and national aspirations shifted to economic development and economic integration between small principalities. This nineteenth-century realism, which had its grand finale in the creation of the German Reich in 1871, was followed by another reaction that again emphasized the cultural and intellectual superiority of Germany. The First World War brought this stage to an end and the Weimar Republic 1919–1933 began life as a faltering, short-lived attempt at economic bourgeois realism, until it, too, succumbed to the National Socialist revival of Kaiser Wilhelm-style megalomania. The Second World War did away with all national feeling; that is, it produced something similar to the effects of 1848, a flight into self-realization through materialism. Today that new sense of identity, based as it is on economic success, is ready to be replaced by a new search for national identity.

That such cycles exist in the history of Germany is plausible, given a situation characterized by both an absence of political institutions and a feverish search for national identity. The institution of Bismarck's Reich was apparently not sufficient to attenuate the cyclic fluctuations. A much longer essay would be required to establish why. Here I only wish to remark that the permanent battle against, and the state of being haunted by, dominant ideas, the sense, almost, that ideas themselves are battling for *Lebensraum*, seems to be one of the more enduring features of German culture, amid all the discontinuity.

The 'rule of ideas' also shows itself in a literary culture, which, contrary to nineteenth-century English and French tradition, and in spite of the dominance of *Realpolitik*, was barely acquainted with the cultural movement of Realism. Before 1871, Martin Swales states, German literature had no counterpart of the St Petersburg of Tolstoy, the Paris of Balzac or the London of Dickens.[18] There was no large metropolis where the vigour of daily city life was exhibited, ready to be absorbed and portrayed by aspiring writers. Even middle-sized German cities were not of a type where agrarian capitalism and its modernizing impulse would cause social discord and nostalgic feelings for a lost *Gemeinschaft* (as occurs in the England of Thomas Hardy). The German cities had had their Golden Age, but in the nineteenth century they had declined into remnants of a feudal society in which human relations were based on old rituals, such as the master–apprentice relationship, rather than on impersonal economic relationships. It was a society which did not offer a tangible alternative for participation in the shape of institutions, life-styles or social groups. This explains why the intelligentsia, longing for change, had no choice but to take refuge in idealism and *Innerlichkeit* (introspection).

The protagonists in this literature are unable to identify themselves with the social order, a fact that forces them to turn to their own inner world or to sterile cultural models from the past. It is not surprising that this literature had difficulty in penetrating to the wider European cultural stage. After Goethe, German literature disappears from the European stage for almost a century, to return gloriously with the Berlin novels of Theodor Fontane (1870). Certainly, introspection was also a feature of the new German literature (Fontane, Mann, Kafka, Hesse, and so on), but it had become a contribution to European realism, rather than a flight from reality. In the same way as realism was closely related with the experience of modernity (and the local as endangered tradition), German culture in the late nineteenth century was the pioneer of post-modernism.

INTO THE TWENTIETH CENTURY

In the event, Germany developed in the second half of the nineteenth century into the most modern state in Europe. There was an awesome growth in industrial production and technological knowledge. On the eve of the First World War, German production of electrical goods was (by value) twice that of Britain and almost ten times that of France. The six largest German firms in the coal-tar industry (important as a supplier of materials for the chemical industry) took out eleven times as many patents as their British counterparts between 1886 and 1900. Berlin was new in its entirety, a metropolis that allowed no comparison with European capitals, only with developments in the New World. As Walther Rathenau said, it was a 'Chicago on the . . . Spree'.[19]

This modern world, in essence the triumph of materialism and rationality, was accompanied by strong feelings of discontent. The German state might have been created, but it appeared to possess no unity or soul, and modernization only reinforced that feeling. However, the cultural and political reaction to this discomfort was not a movement with one unequivocal programme. The cyclical vision of German history logically stresses the anti-pragmatic thrust of this counter-movement, but so many paths had been traced since the inception of the German state that one has to account for divergent impulses. Here, Marx, May and Wagner have been evoked to symbolize those divergent paths. This is not to say that their messages were important in themselves. They only represent the kinds of input that made up the spirit of the times.

With his materialist perspective and emphasis on 'inevitability', Marx seems to belong to the domain of Bismarckian *Realpolitik* rather than to the cultural acts and mystical confessions which Wagner and May advance as the road to national salvation. However, Marx does represent a broad view over space and time that matched a need to transcend German 'local patriotism'. Already in 1848, Marx and Engel's *Communist Manifesto* prophesied: 'National one-sidedness and narrow-mindedness will turn out to be more and more untenable, and from all the national and local literatures will arise one world-literature.' What country would fulfil the function of epicentre for such movements better than Germany? The idea that the world might integrate and ultimately disintegrate into a limited number of blocs – world empires – was already being discussed in Germany in 1890. Gustav Schmoller foresaw the genesis of Russian, British, American and Chinese empires. Alongside these there seemed to be room only for a *Mitteleuropa* (Central Europe) under German guidance.[20] Such thoughts did not make use of Marxist theory but they shared with it an international perspective and an assumption of inevitable social change.

Karl May, the moralist, also displays a cosmopolitan vision, with his stories about North America and Kurdistan. Fundamentally, he agrees with Marx's conception of the material determination of human consciousness, albeit in a way Marx never imagined. Those who come into contact with gold will display symptoms of fever, a trial hardly to be borne, even by a German hero. Old Shatterhand only succeeds because he is subjected to the purifying force of the primitive life-world and its physical environment.[21] However, behind such high-toned stories lies the petty-bourgeois fear of loss of personal decorum, of moral 'devaluation' under the sway of commercialization. This is the kind of unsettled state that, according to all classic texts on nationalism, intensifies the susceptibility to nationalism because only national identity can transcend sudden changes in prosperity while at the same time providing visions of a lasting superiority.

Among these intellectuals Richard Wagner is the only one who puts the human will, not the environment, first. When Franz Liszt, Wagner's father-in-law, inspects the construction site of the musical theatre in Bayreuth (1873), he exclaims that for the first time in history a theatre is being built for an idea (namely the performance of Wagner's operas). Replace 'theatre' by 'state', and we are in the new Germany, a state which was easier to understand as a result of the human will than as an expression of geography or ethnicity. In Wagner's view the state is not complete without Idea, without 'Art'. When the *Festspiele* (the Bayreuth festival) finally took place and the public had cheered the *Götterdämmerung*, Wagner climbed the stage and said: 'Now you have seen what we can do, do you also want [. . . *willen Sie auch*] ? And if you want, then we have an Art!' Wagner would prefer to destroy this entire world of institutions with all its material products, and start afresh with a *Gesamtkunstwerk* (work of total art). An unequivocal and, to the contemporary ear, uncivilized destructive urge emanates from his autobiography and from the diaries of his wife Cosima. After the events of 1870–1871 Wagner called war 'a sublime event that expresses the insignificance of the individual . . . , a ceremonial dance with the dreadful, like a Beethoven finale in which the composer unleashes all the demons in a splendid dance.'[22]

What can we make of such contradictory attitudes and statements? The turn of the century produced a mix of cosmopolitan thinking and xenophobic nationalism, a sense of inferiority and megalomania, idealization of primeval habits, contempt for 'lower' civilizations, a belief in personal genius, and the enjoyment of being lost in the crowd or in an orgy of destruction. Relating such visions to rapid economic growth and to the lack of time-honoured political certainties seems an obvious answer. The situation in late nineteenth-century Germany inspired both the intellectual ability to transcend the limited world of daily experience and feelings of anxiety at the same time. The existence of high-toned talk about life, separated from the world of practical problems, was nothing new; the split simply continued after 1871 with high-level discourse shifting from the mythological and poetical to (quasi-)science, high Art and new humanities. To judge from the elite visions of around the turn of the century, a Germany which could have prompted an early twentieth-century European Union was a possibility. However, the nationalist chemistry of the times decided otherwise, a fact that was undoubtedly not solely due to shortcomings on the side of Germany.

German idealism did not prevent the building of a German war-machine, nor did the theoretical praise of pure and natural ways of living lead to a positive appreciation of the 'cultural-fall' in an eastward direction. On the contrary, the visionary geography of Marx and Darwinist thinking became 'translated' into the idea that other peoples and states were resisting the inevitable in history, and that the forms of society 'beyond' resulted from degeneration rather than from natural principles.[23] Instead, primacy was looked for in the German past, in the colonization activities of the Teutonic Order 700 years earlier. The praise of foreign examples in previous decades had become too much for the German people.

For Heinrich von Treitschke and other historians in his wake, the Teutonic

Order and its penetration in the barbaric Prussian, Polish and Latvian regions was the symbol of German primacy. With an almost malicious satisfaction Treitschke brushes aside illusions about a friendly transfer of culture: there was much blood shed in this 'racial war' (1862). 'It was genocide, it cannot be denied,' Treitschke later serenely observes, but denying was precisely what an objective analysis would have required.[24] This representation was not sanctioned by historical facts. Thus we see emerging the bizarre picture of an historiography that aims to emphasize the inhuman in the country's past, an attempt to counterfeit reality in the name of realism (the *jenseits von Gut und Böse*, beyond Good and Evil).

This representation of the past did not necessarily mean that historians at the end of the century were aiming at a new politics of expansion to the east. They were concerned, on the one hand, with the creation of an identity, building a myth about the vital power and willpower residing in the German people. On the other hand, they tried to practice scientific 'realism', showing that the world, with all its small, time-bound human worries, obscures a view of the cold processes of demographical selection and large geographical movements. It was not at all necessary to invoke the image of military violence. The confrontation between German and Slavic forms of life within the same territory would automatically lead to a more or less peaceful 'extermination' of the Slavic element, by cultural assimilation into the superior German race and because the Polish would not be able to compete economically with the German activities, a mechanism that would gradually transfer land into German hands.

However little these authors were appealing to military solutions, their conceptions matched very well the aggressive national politics that the National Socialists championed. It is not surprising that historical and geographical studies in Germany, even after the First World War, continued to spread this body of ideas. Social scientists discovered a splendid opportunity to prove the social relevance of their fields of study and there was no trace of critical comment on this servitude to power in Germany between the two world wars.[25]

NATIONAL SOCIALIST PHANTOM-LIMB PAIN

After so much shame falling upon our country, acknowledgment of 'Civis germanus sum' must be repeated proudly and emphatically. First, the rising generation has to have spiritual contact with the violently separated parts of the German territory and with their populations. Each classroom should have a map which shows the old border, so that it is obvious at a glance what has been lost in the First World War.

(F. Schnass, *Lehren und Lernen, Schaffen und Schauen in der Erdkunde*, 1923[26])

In 1933 the Polish geographer Kazimiera Jezowa published the results of an analysis of the content of German geographical journals since the First World War. Each of the 54 volumes analysed contained at least one anti-Polish article.[27] These polemical writings always seemed to focus on two themes: the 'so-called rights of Prussia to the territories occupied by Poland' and 'the so-called inferiority of the Polish people'. At the International Geographical Congress in Amsterdam (1938), Jezowa admonished her German colleagues to maintain scientific standards of reasoning. She did not get them to repent; on the contrary, they heatedly debated between themselves about how German geography had to confront the Poles.

After the First World War the preoccupation of Germany with its geographical surroundings reached a new culmination. The territorial amputations enforced by the 1919 Treaty of Versailles, the emergence of the Polish state, which had been wiped off the map for more than a century, the appearance of other small East European states, and the apparent crisis of the liberal market system (revealed in the Crash of 1929) were all incentives to reject the existing world order and to look for the profile of a more profound historical process and conditions which would do more justice to the German 'essence'. Researchers who identified with this attitude produced studies at two spatial levels: those of *Ostforschung*, a generic term for regional studies of lands to the east, with a cultural, rural-sociological, historical or geographical subject-matter; and those of the more European or globally aimed studies of *Geopolitik*, the notorious school of Karl Haushofer (1869–1946). In both fields the concept of *Raum* played an almost mythic role.

The historian Hermann Aubin observed in 1937 that the territorial losses after the war had forced German historians to transcend the limitations of 'dynastic-territorial historiography'. This means that historians, in focusing on regional subjects with a strong ethnological and cultural-geographical character, had started to overstep political boundaries – an approach which, with hindsight and the knowledge of Braudel's innovation in historiography, might even be called 'modern'. But the message was tendentious.[28] All theory was based on the aim of verifying the 'unhealthy' or 'un organic' settlement structure of the Poles, the 'irrational' economy in the Slavic territories and the fact that even where no Germans lived, traces could be found of more ancient German civilization. Were such facts supposed to support territorial claims? Explicitly no connection was usually made, but even tacit facts could help the plea to interpret German aggression as self-defence. This impression was even reinforced by the westward direction of demographical movement within Germany. However, this internal migration indicated the pull of the economically highly developed regions in the west rather than any pushing forward by the Slavs.

Whereas supra national thought around the turn of the century manifested itself mainly in speculation about economic territorial expansion and in high-toned statements about the salutary influence of the colonial experience on the German character, the frustrations of the war induced an almost physical experience of political space, a phantom-limb pain.[29] Even Walter Christaller (1893–1969) – later widely accepted in Anglo-Saxon geography as the originator of 'value-free' spatial economic models, even though he placed himself unquestioningly at the service of the Nazis – assumed the emergence of a special 'German awareness of space' since the war.[30] The picture of a people which was bursting at the seams in its allotted space was reinforced in every possible way. Geopoliticians juggled with maps and figures to prove that Germany was the most densely populated country in Europe by adding the colonies to the territories of Britain, the Netherlands and France, or by incorporating all uninhabited and uninhabitable Siberian and arctic regions in the figure for Russian population density.[31] If we add the fact of a German-speaking nation that is geographically distributed far beyond the political boundaries of the German state, then the picture of an artificially and unfairly constricted country is complete. The antipode of this picture was the boundless space of the East, a territory that was believed by the Nazis to be almost empty and ready for a new millennial colonial enterprise. An official document from the early years of the Second World War (1941) paints the following picture:

> The enlargement of the German sphere of influence across large parts of the Russian territory will be of greater revolutionary importance to European culture than any other event since the invasion of the Arabs in the Mediterranean basin. In view of such fundamental processes, not only will the German awareness of space, fixed for over a millennium, be revolutionized, but also social thought in general.[32]

The actual German push in Europe under Hitler did not proceed according to the deeper lines of *Schicksal* (destiny) that had been discovered and traced by German geopoliticians in the period between the wars. Although these writers feared the machinations of France and hated French culture, their geopolitical theories did not stress the importance of French, Belgian or even of Dutch territory.[33] Their gaze centred on the east and southeast. There, in the area where in medieval times the first Reich (the Holy Roman Empire), too, had been situated, was German destiny to be found. That destiny was *Mitteleuropa*, an area which contained at least the Bismarckian Reich and the countries of the old Habsburg monarchy, the Danube basin. The geopoliticians conceived of such an empire as the fulfilment of fundamental but veiled geographical conditions to which a new economic meaning was added. The cumbersome development of international trade seemed to forecast the failure of world trade capitalism. The new economic world order would tend towards the development of a number of autarkic world regions, such as an American-British and a Japanese-Asiatic empire. *Mitteleuropa* was a logical third such empire, and the geopoliticians had already established that this area would be self-sufficient in products such as grain.

The area of medieval German colonization, the territory over which the German-speaking tribes had swarmed out over the course of time, was, according to the geopoliticians, a better indicator of 'immanent geopolitical tendencies' (today we would say the tendency of self-organization) than the state territory of 1871. However, this natural unit extended over all of Eastern Europe, even beyond the Dnieper. On the question whether this had to be one state or a confederation of independent territories under German guidance, no unequivocal answer was given. In any case, the Rhine, Danube and Weichsel (Vistula) determined German *Schicksal*.[34] These three rivers were lifelines in an economic, geographical and also in a spiritual sense. Their basins were the natural domain of an 'eastern people' with its own psychic and cultural characteristics deviating from the western type of culture. Collective and instinctive qualities made this culture a negation of the principles of the French Revolution – principles such as the right of self-determination of nations or the democratic rights of individuals – and of rationalism. Everyone, either as individual or as minority, would have an organic place within this Reich and everyone would benefit if they acquiesced in that role.

The idea of a predestined *Mitteleuropa* that would manifest itself to the world by means of self-organization or 'self-determination' (just as Germany itself had done) would, in the geopoliticians' view, have horrible consequences when it was thwarted. If the Great Powers nonetheless tried to enforce their own artificial rules and limitations, then this natural process could only end up in a great World Fire in which Germany would go down together with all of Europe. As later became clear, the Nazis were on their way to converting this vision into a self-fulfilling prophecy.

Geopolitical thought in the 1930s and 1940s cannot be equated with either German public opinion or the aims of Adolf Hitler and his inner circle. The new borders after the First World War were almost universally experienced as unjust, because the German public assumed a natural right to the unity of state and people (including the Austrians). The visions of the geopoliticians extended to more distant horizons, albeit not as far as Hitler wanted to go. For Haushofer and others, war with the West served no sensible goal. The German–Russian non-aggression treaty of 1939 (the Hitler–Stalin Pact) was more Haushofer's line because it would bring the goals of *Mitteleuropa* nearer. Whether Hitler's plans were more expansive from the beginning remains open to question. To view events between 1939 and 1945 as the autonomous dynamics of war and of the military mode of production is probably a more rewarding approach than to search for an initial master plan. Hitler recognized no borders, which is certainly not what geopolitics teaches.

'SYSTEM-OPENING COEXISTENCE'

Germany's relationship with Russia has always been ambiguous. In geopolitics fear of and fascination with Russia went hand in hand. Geopoliticians like Obst and Haushofer did not like the general character of the Soviet system but they appreciated the ostentatious rejection of the liberal, Anglo-Saxon body of thought. Both communist agitation in the colonies and the supposed mystical affinity with the Russian national spirit evoked a certain understanding.[35] And, of course, there was the message of Halford Mackinder, who attributed an important role to the Eurasian core as a pivot of history and, later, to Eastern Europe as the key to this pivotal area. Immediately after the First World War, Mackinder had already pointed out the mistake the Germans had made by wasting their energy at the Western front. Without the exhaustion caused by these war efforts, a treaty might have been forced out of the Russians, which would undoubtedly have been a surer road to world power.

After the Second World War, the Federal Republic of Germany (FRG), although firmly dedicated to democratic principles, was the most active member of the North Atlantic Treaty Organization in creating and maintaining a dialogue with the Soviet Union and Eastern Europe. On the face of it this might seem odd, because the Russian occupation of the Eastern Zone (later the German Democratic Republic, GDR) brought an alien system within reach. But it was precisely this division of Germany that incited the *Ostpolitik*, a rapprochement with the East that was carried out over several decades with the patience of a saint. Hope of reunification was never given up in either of the two Germanies. The possibility was ratified in a number of treaties between the GDR and the Soviet Union, until it ceased to be mentioned in 1974 (but was not denied either). [36] As time progressed, an end to the division of Germany and of Europe became more and more difficult to imagine, and looked farther away than ever. The Western allies and Germany's European neighbours displayed little enthusiasm for such overtures and the Soviet Union tried to hold on to this trump card against the West as long as possible. This explains why in Germany both public and politicians left such aims out of the discussion, or even started to attach a positive value to the status quo as a way of exorcizing the evils of the past. [37] Even in the 1980s political commentators observed that the FRG had no effective means at its disposal to facilitate German reunification. [38] The only possibility was the passive aim of keeping all future options open, of taking no irreversible decisions. That counsel determined the actions of all important politicians in the FRG, so that they continued to state – even under the most favourable circumstances, as Willy Brandt still did in 1989 – that reunification remained an illusion for the foreseeable future.

There were more active options too, although they were related very indirectly to the goal of German unity. These options can be subsumed under the label *Ostpolitik*, a strategy that in its broad outlines was implemented in a similar way by successive federal governments. The relevant actions, most directly pertaining to the GDR, primarily implied an attempt to keep communications open between both Germanies, pleading the case for the human rights and liberties of GDR citizens, alleviating their living conditions, and guaranteeing free access for FRG citizens. In addition to genuine concern for the fate of *die Menschen* beyond the (zone-) border (who were often close relatives), they tried to project a positive image of the FRG. If some day the option of self-determination for the German people should become a reality, a right which was never denied, then this would favour the likelihood of the people 'over there' (*drüben*) choosing to join the FRG. When that point was in fact reached in 1989 the West Germans could with some self-satisfaction speak about *Beitritt* (accession) by the GDR. Throughout this entire process, an urban authority had taken the lead: Berlin, under the direction of Willy Brandt. What had first been urban later became national politics, and even acquired, as a result of the East European *Wende* (upheaval), international dimension.

The 'model country' strategy also dominated foreign policy towards the other East European countries, but here human rights and individual conditions of life were much less emphasized. As Timothy Garton Ash has exhaustively illustrated, the catchword of German *Ostpolitik* was 'stabiliza-tion', not 'liberalization'.[39] This policy aimed in the first instance at inspiring confidence in the East European leaders, to reassure them regarding German intentions. It was a statement against everything the Third Reich had meant to Eastern Europe, involving solemn declarations that never again would a war start from German territory, that the FRG did not have nationalistic aspirations or vengeful feelings, that economic collaboration was the only foreign policy aim, and so on. These messages were reinforced in domestic policy: an exemplary functioning of democracy and a generous asylum policy. However, this approach left no room for supporting dissident movements or for a firm political response to East European acts of state terror. The remarkably moderate reaction of Helmut Schmidt to the Soviet invasion in Afghanistan (1979) was that 'the politics of détente would not survive *another* event like that'. It was one of the many political statements that laid bare the political difference between the FRG and the United States. Risking the politics of *détente* was precisely what the Americans were willing to do, and indeed did, during the first years of Ronald Reagan's presidency. The United States did not shrink from confrontations with Eastern Europe about human rights, indeed 'destabilizing' communist regimes was almost their supreme goal. However, taking into account the geostrategic context, there is a great similarity between the political choices of the United States and the FRG for or against destabilization and defence of human rights. Commitment to human rights has always been very selectively raised in the United States.

When the Polish workers revolted under the banner of Solidarity and martial law was subsequently proclaimed (13 December 1981), the first reaction of Helmut Schmidt was to express the mutual distress of himself and his GDR colleague Honecker about the 'necessity' of this development. This fraternal association with the GDR regime (which, of course, praised the *coup d'état* of General Jaruzelski) was partly disavowed in the days that followed, but initial reactions are often more telling about visions and feelings than dispassionate diplomatic messages. The Americans organized a monu-mental TV show – a 'hysteria' according to Schmidt[40] – to support the Polish people morally. In Germany nothing happened. Herbert Wehner, the FRG's minister for 'intra-German relations' who worked himself up from the German workers' movement, travelled to Poland as early as 1982 for a cordial meeting with Jaruzelski. At the same time he was distributing among his parliamentary colleagues a call for support of the Bolivian 'consulate in resistance' in the FRG.[41] In America the feelings of solidarity were precisely reversed as became very clear after the murder in 1984 of the Roman Catholic priest Popieluszko by the Polish secret service. As a content analysis of news items by Herman and Chomsky has revealed, interest in the case was overwhelming in America. In the same year, however, when some American nuns were raped and murdered by secret troops of the El Salvador army, US reporting of these facts differed strikingly from the news of the Polish murder, both in quality and quantity. American reports on the events in Central America even put part of the blame on the nuns themselves because they had 'meddled with Salvadorian politics'.[42]

In any case, the FRG had one good reason to be reluctant in supporting dissident movements: the national past. The government tried at any price to prevent Germany again assuming the distinctive character of an intriguer, an intruder bringing dissension in Europe. It was hoped that inspiring the confidence of the East European leaders would be the best way to liberalize these countries, and that approach tended to put aside any painful moral question. But was it a sound argument? Garton Ash states: 'The West could never provide enough assurance to make communist rulers relax, because the internal tensions came from the very nature of their regimes and not merely from the external tensions of the Cold War.'[43] Unconditional Western support, irrespective of domestic policy, would not be interpreted by the communist leaders as an incentive to liberalize but rather as evidence that their hard line was rewarding. The Federal German line of 'system-opening coexistence', of liberalizing by means of stabilizing, was bound to fail. The policy was never understood by members of movements like Charter '77 in Czechoslovakia and Solidarity in Poland, who risked their freedom by pleading publicly for political change. These dissidents were disappointed by the lack of German support, even from people like Willy Brandt who knew what it meant to be suffocated by a totalitarian regime. The Federal German government's dilemma was that a policy aiming to change East European societies, at the same time alienated members of these societies from Germany. Ultimately, resistance from below turned out to be the deciding factor in the ending of communism, not the fraternal message of *détente* to the leaders. There was of course also an autonomous erosion of confidence in the communist system caused by such factors as the technological gap with the capitalist countries (in computers), environmental problems (as shown at Chernobyl) and the comparative success of underdeveloped countries that had strayed from the communist doctrine (for example, China).

None of these decisive factors was a consequence of specific German policy. If political action contributed anything at all, only the United States may be considered as an influence, because, contrary to the FRG, it did not recoil from supporting dissident movements. However, although German *Ostpolitik* probably contributed little to the East European collapse as such, Garton Ash suggests that it may nevertheless have been productive for the FRG. It was precisely through the politics of restraint, through being handcuffed in golden chains by its European neighbours, that the FRG created a climate allowing the maintenance of relations with the new democracies to the east that were more relaxed, less burdened by the past than could have been possible after the preceding, more assertive, style of politics.

AFTER 'DIE WENDE'

In spite of these positive developments, it is unlikely that this German propensity for self-restraint, this willingness to go through life in golden hand-cuffs, will last forever. In this respect Germany's restraint is too rooted in its exceptional history. The new generation of politicians and intellectuals, of whom hardly any have a personal memory of the war, and some of whose parents were indeed barely politically conscious, no longer accepts a special position for Germany. The resistance to the legacy of the war has slowly been enhancing its intellectual framework since the beginning of the 1980s, and the advent of a more positive attitude to the national identity may fit the cycle of reversals that can be observed throughout German history (see page 23).

Postwar introspection in the FRG relied heavily on the deep-rooted thesis of a German *Sonderweg* (an exceptional course through history). In the nineteenth century the idea of exceptional development was still positively valued. The German maturation as a 'culture-nation' without an overall state apparatus was seen as something unique and superior. This idea was even invoked by foreign commentators. After the Second World War the meaning of *Sonderweg* became diametrically reversed and the term came to denote a pathological deviation from the 'normal' development that other countries are supposed to follow. The Western alliance was the tested remedy against relapsing into the *Sonderweg* and its excesses.

The current level of intellectual resistance to this point of view can be inferred from two collections of essays published in 1993 and 1994. The contributions to one of these volumes, appearing under the title *Westbindung* (Being tied to the West), can be summarized in three arguments.[44] First, the *Sonderweg* thesis is disputed: 'normal' historical developments of countries do not exist and consequently exceptional developments do not either. Second, the German alliance with the West is pictured as an almost religious devotion which prevents a pragmatic view of Germany's external relations. This holds true particularly for German enthusiasm for further integration into the European Union. The message is that Germany's unselfishness is not completely requited by the other EU members. German politicians should be aware that France has always preserved certain geopolitical aims that it boldly pursues, such as closer association of states bordering the Mediterranean with the EU. Germany should learn from this to promote her Central European interests more boldly. Third, the intended reinforcement of national consciousness does not presuppose the political ideal of a 'third road'. The authors do not want to denounce the ideals of democracy and human rights or to separate Germany from the West in general. That is, at least, the reassuring note in this shift to a more nationalistic stance.

A more conservative and aggrieved note is sounded in *Die selbstbewusste Nation* (The Self-conscious Nation). Here one meets a hardly veiled hostility, particularly directed at those left-wing intellectuals who have talked the Germans into a collective feeling of guilt. 'Long before the concept of *political correctness* was coined in American East Coast universities, we were coerced into left-liberal attitudes, into taboos on themes and groups of persons, the observance of which was watched over by a thought-police, as in the USA . . .'.[45] In order to delineate the differences between Germany and its Western allies, geographical concepts like 'middle', 'the German's ties to the earth', and even the 'eastern value system', assume, once again, an almost mystical value in this literature. Western thinking and cosmopolitanism is imputed to have transformed the Germans into displaced persons. This is uniquely disturbing because there is no European people (the Russians perhaps excluded) which so profoundly loves its home country and region of birth (*Heimat-liebe*) as do the Germans, a feeling which is even intensified because they are a 'metaphysically homeless people' (Nietzsche said 'the German . . . is not, he *becomes*'). Again the simplicity of Germany's postwar international posture is castigated: 'People without national identity are not pioneers of a new world order, but victims of foreign interests.'[46] Against those who want to repress German national identity the argument is advanced that this option openly clashes with other highly esteemed political attitudes, such as the support of 'national' liberation movements elsewhere in the world. The conclusion is that Germany should lose its fear of thinking in geopolitical terms and stop its habit of dealing with supranational bodies like the United Nations or NATO as if they were sacrosanct institutions.

These 'normal nationalists',[47] as they are sometimes called in Germany, should not be equated with public opinion or with tangible influences on the perception of the foreign policy elite. A content analysis of foreign policy documents has shown that the dominant trend still is to assign Germany first, to maritime Europe, west of the Baltic-Adriatic isthmus,[48] in brief to the territory of the EU, and, second, to the northern hemisphere generally (United States, Commonwealth of Independent States (CIS). With the exception of the Middle East, the rest of the world is strikingly missing in Germany's 'geovision'. Other 'natural frontier' in Europe, such as the line between the Baltic and Black Seas, or the Urals, do not appear to have much significance in foreign policy documents and aims. The East is a continuous zone, an atmosphere becoming thinner and thinner until it ends at the Pacific. It is improbable that Germany as a 'country of the middle' will remain with its back turned to the east; but this geopolitical vision has yet to mature, as the neo-nationalists claim themselves. Is this how matters will develop?

There are some reasons why the nationalist perspective might become more widely accepted, on account of the logic of some of its arguments and the inevitable erosion of the legacy of the Second World War. First, there is a profound truth in the conviction of the Right, that a country cannot face the future with a purely negative motivation. All postwar counsels for Germany's politics have been based on *cutting off* options, inclinations and memories. The concept of *constitutional patriotism* (*Verfassungspatriotismus*) in place of the nationalistic patriotism of the Nazis is possibly insufficient to create a sense of meaning and belonging. Social policies focused solely on repressing behaviours have never turned out to be very successful. The chances of success are better if policies involve positive options, goals and rewards.

Second, the demand for tolerance of the nationalist aims and selfishness of other peoples, while any particular feelings of German-ness and promotion of national interests have to be repressed at home, cannot be seen as a coherent position. Although the Germans have little to complain of in their current predicament, there are facts which can easily confirm this view in the public mind, such as the absence of a German seat in the Security Council of the UN and the financial burdens connected with the Deutsche Mark as a stabilizing currency.

Neo-nationalist writings demonstrate that geopolitical visions present in German political and intellectual debate at the end of the nineteenth century, for example in the writings of Friedrich Ratzel, have not lost their attraction. The problem of Germany's many neighbours, its situation in 'the middle' of Europe, the instability of the East, even the special German attachment to the land, are back again. This does not mean that these ideas will get the same continuity nor that they will attain the same power and meaning as during the first half of the twentieth century. Indeed, although certain writers currently feel a need to rehabilitate Karl Haushofer to some degree,[49] or to evoke once again old philosophical doubts about the loss of human dignity in mass consumer society,[50] the basic aim is still a democratic society, economic rather than military power, introspection and stability rather than a world-wide mission. Geopolitics owes its revival in Germany to the shattering impact of the Second World War. It is a correction of another correction to the swing of the pendulum, with all the provocative intellectual notes that we can expect to hear particularly in Germany.

Those who have written about the notorious German *Angst* have somewhat weakened the value of this concept by pointing out so many sources for this feeling. Both geography and the repercussions of the last war are mentioned as causes. *Angst* may be a profound feature of cultural psychology (as in Nietzsche's 'becoming') or just a dramatic way of describing the post-war political situation of Germany (as it was used by Helmut Schmidt). Although political conditions have improved drastically for the Germans since 1989, reminiscences of the preceding period may continue to have their psychological effect for some time. The longing for stability (the reverse of *Angst*) has been identified as one of the most pervasive concepts in German political languages in the 1990s.[51] The interpretation of *Angst* as something of the *longue durée*, evoked either by 'permanent' features of Germany's geography or by education, is a more speculative thought. It is certainly remarkable how much the intangibile East has been a continuous factor in Germany's geopolitical experience. This experience has not been interrupted by the post-war period, indeed the division of Germany even made it the most serious existential threat in its history. Even after reunification Germany cannot allow itself to ignore Eastern Europe as a source of instability. The need to cope with the changes in the modern world and to guard its own interests in the Western alliance will in all probability in the near future reveal a Germany that is more 'national' than its neighbours have become accustomed to over recent decades.

ABSENT BECAUSE OF EMPIRE

(Britain)

THE BATTLE OF DORKING

'Sind wackere Soldaten, diese Englischen freiwilligen,' said a broad-shouldered brute, stuffing a great hunch of beef into his mouth with a silver fork, an implement I would think he must have been using for the first time in his life.

'Ja, ja,' replied a comrade, who was lolling back in his chair with a pair of very dirty legs on his table, and one of poor Travers's best cigars in his mouth; 'Sie so gut laufen können.'

'Ja wohl,' responded the first speaker; 'aber sind nicht eben so schnell wie die Französischen Mobloten.'

'Gewiss,' grunted a hulking lout from the floor, leaning on his elbow, and sending out a cloud of smoke from his ugly jaws; 'und da sind hier etwa gute Schützen.'

'Hast recht, lange Peter,' answered number one; 'wann die Schurken so gut exerciren wie schützen könnten, so wären wir heute nicht hier!'[1]

And so the English were humiliated after the Battle of Dorking early in the 1870s. The company of German occupiers was watched by a wounded 'volunteer' reporting fifty years later about this dawn of English misery as a lesson to his grandchildren. The German soldiers appreciated the bravery and shooting skills of the English volunteers, but lack of organization and discipline had finally decided the fall of Britain.

It was all fake, of course, this short story written by Colonel George Tomkyns Chesney and published in 1871. For a long time after its publication the question 'Weren't you wounded at the battle of Dorking?' could raise a laugh. But timely publication of the story, just after the German entry into Paris, made a tremendous impression. It started a boom in invasion stories in following decades, lasting until the eve of the First World War. The genre appears to have been dominated by the English, who to a limited degree infected the French. The latter did not reverse the invasion's direction but preferred nationalistic fantasies about their own joyous Channel crossing. Self-torturing fantasies were more typical of the British, which may say something possibly about their scepticism about 'grand designs' but more probably about special fears connected with the insular existence, not least the late nineteenth-century fear of 'imperial overstretch'.

The geography of the British state evoked not only fears of invasion but also the assessment that occupation or defeat would be more disastrous than in the case of a continental power like France. As the narrator of *The Battle of Dorking* relates, 'France rose again.' Its wealth depended on a country(side) that was rich in itself. But Britain's wealth depended on commerce and colonies, all of which were swept away by the defeat at Dorking. The British Empire ceased to exist and the story depicted a rump nation with bleak prospects.

The appearance of the late nineteenth-century 'future war novel' was certainly not the start of British invasion *fears*.[2] Apart from reflecting events in France, it basically marked the breakthrough of a new popularizing medium, new audiences and new ways of publishing. Invasion fears themselves were at least as old as the French Revolution and the Napoleonic military expeditions. According to Linda Colley:

> We know from diaries of the time just how common nightmares of a violent French invasion of Britain were at the end of the eighteenth and the beginning of the nineteenth centuries, and how easily rumours that the enemy had in fact landed gained ground among the credulous and the frightened. . . . Nerves were kept at fever pitch by preparations for evacuation, by long lines of wagons in village streets waiting to transport women, children and the infirm away from the scene of battle . . .[3]

Such fears and the preparations to avert a French attack were actually part of the process which Colley analyses as 'forging the [British] nation', a process of national identification based on three pillars: Protestantism, the empire, and war. The volunteer corps beaten in the Battle of Dorking had been created as a response to the French revolutionary armies. They were deemed necessary in order to match the military strength of a belligerent neighbour, but they also accelerated the process of national unification. It was a piece of revolution imported into Britain, in spite of all the negative judgements on the militaristic nation across the Channel. For the first time the ordinary people were trusted with weapons without the fear that they would be used against the domestic ruling class. A common national feeling was born out of an appropriate enemy abroad. France provided the most articulate 'foreign' image, with all its connotations of danger, throughout the nineteenth century, only to be supplanted by Germany in the first decade of the twentieth.

In the course of time the perspective of the literary-invasion genre in England shifted – it survived both world wars better there than it did on the continental mainland. Spiering, discussing novels by Angus Wilson, Kingsley Amis and Anthony Burgess published between 1961 and 1980, establishes an important distinction between early and late invasion novels:

> The central difference between the two categories is that in the early stories England, as we have seen, is ill, but there is every possibility that she shall recover. The fear has an incidental nature; a 'little firmness and common sense', to quote again Chesney, and all shall be well. The fear expressed by late invasion stories, on the other hand, is an existential

one. In these stories the question is not whether England will recover in time, but whether there is any England to recover. . . . England is not just sick; her brain has softened. No medication could induce recovery.[4]

Comparing *The Battle of Dorking* with Kingsley Amis's *Russian Hide and Seek* (1980), which also opens fifty years after an invasion and the subsequent occupation, Spiering concludes that in the later stories the focus of anxiety has shifted from fear of destruction to fear of dissipation of nationality.

The invaders in the early stories descend on England as predators on their prey. In the later stories they appear to be sucked in as by a vacuum; now the question is not whether the English nation will conquer or be conquered, but whether it will survive.[5]

These were the fears so aptly voiced and converted into politics by Prime Minister Margaret Thatcher and Anti-Common Market demonstrators in her wake. The nations outside – the European Union now, the Americans previously – are seen as the great suppressors of variety and identity or (in the case of the continental powers) as the bearers of bureaucracy and absolutism. The fear of losing one's identity in the current process of homogenization and international exchange, however, is by no means exceptionally British. In other European countries a revival of the identity question can be established, with the same basic motives. As Perry Anderson has argued, a subjective desire to (re)construct an identity where a social and historical underpinning is inevitably failing even pervades such scholarly enterprises as Braudel's *The Identity of France*.[6] The British share a concern about transnational threats to their identity with their European neighbours, but they also have to cope with traumatic geopolitical changes.

THE GEOPOLITICAL HERITAGE OF THE NINETEENTH CENTURY

The Battle of Dorking may have touched in the British public the right chord about national security, but the Foreign Office was not pleased. In a speech Gladstone, referring to the book and the stir it had caused, warned against the dangers of alarmism, in particular with regard to the impression such discussions would give in other countries. Apart from the improper display of self-doubt and the resulting loss of 'credibility' for the British, the novel had proclaimed the Germans to be the next enemy and this was obviously at variance with long-standing British foreign policy aims.

British world politics at the time was primarily engaged in questions about the security of the empire, whereas policy-making towards neighbouring states espoused the 'free hand' principle. This implied a belief in the 'natural' mechanism – later so admired by Henry Kissinger – of a balance of power in Europe but also a refusal to enter into alliances with one or more of the other European powers. In fact British policy was not geared to waging of war at all: indeed, the prospect of a military victory frequently evoked dismal comments on the costs of such ventures, either to the nation in terms of military operations or to the enemy in the shape of disruption to foreign markets.[7] The empire had been held together with gentle pressures and agreements based on mutual advantage, and on the spontaneous attraction of the products of industrial capitalism and its enlightened administration. The hope that such conditions would continue was matched by a belief that wars would be unnecessary in future as each European state would follow the same path of development as that on which the British were the most advanced. It is true that the emergence of the German state in 1871 changed the balance of power, but to draw conclusions about British security in public fitted ill with foreign policy tradition. Britain could cope with its familiar enemy France, but the emergence of new threats evoked the frightening image of inevitable, and unholy, alliances. Moreover the immediate threat after German unification was too small to justify the social disruption which sounder security strategy would entail: higher taxes, military conscription, secret police, and so on. As Bernard Porter has argued, isolationism and liberalism had become too much ingrained in the social system to allow each separate change in the international environment sufficient weight to overthrow the prevailing detached mode of foreign policy making.[8] On the other hand, taking all such events together boded ill for the future of an unaltered nineteenth-century mode of foreign policy making.

The geopolitical vision of the Foreign Office in the nineteenth century emphasized a zone of vulnerability connecting Britain with its Asian colonies: from the Strait of Gibraltar through the Mediterranean Sea to the Middle East. Here British sea power had always been decisive, but only on condition that the continental periphery was occupied by weak powers like Egypt, or other parts of the Ottoman Empire, willing to offer (or unable to refuse) ports and military bases to the British. These comfortable conditions would soon deteriorate as other colonial or naval powers, such as France and Russia, became stronger. The opening sentence of Bartlett's book on twentieth-century British foreign policy summarizes the end of nineteenth-century conditions in one statement: 'At Fashoda on the Upper Nile in 1898 the British achieved their last relatively inexpensive major success in foreign policy.'[9] The 'Fashoda Incident' brought Britain and France to the brink of war, but France, weakened by domestic political turmoil (the Dreyfus affair), ultimately yielded.

After the fall of Napoleon, Russia had remained Britain's only significant European opponent, threatening alternately the Mediterranean sea routes and India. At least, so it appeared on the British mental map. One of the first pamphlets to sound the alarm, published in 1817 and written by Sir Robert Wilson, drew attention to the fact that 'England devoted all her resources to remove the danger of one domineering rival, France; but Russia profiting by the occasion, mounted to a higher pinnacle than that rival ever reached.'[10] Although 'public' opinion was initially divided between alarmists and non-alarmists – the latter appealing to the strength of Britain's industry and navy – Russophobia gained strength over the decades that followed. Events like the introduction of high protective tariffs by the Russian government, Russia's cruel treatment of the Polish people, the advantageous treaties made by

Russia with the Ottoman rulers making Turkey almost a protectorate, and Persian military expeditions supported by the Russians, all contributed to the British image of Russian foreign policy as an ambitious imperialist project.

Gleason has meticulously elucidated how the distrust of Russia originated in early nineteenth-century Britain and who were responsible for the introduction and power of the visions lying behind it. He found a discord between, on the one hand, certain opinion leaders, writers of brochures, and newspaper articles, and, on the other, those practising statecraft. Responsible ministers like Lord Palmerston resisted calls for a more hostile policy toward Russia. By competent manoeuvring Palmerston managed both to maintain a low degree of tension between both countries and to produce a decrease in the output of anti-Russian propaganda. But this was at most an exorcism of merely the most aggressive outpourings. The visions of Russia were merely hibernating until the next crisis announced itself. When this came in the shape of the Crimean War (1853–1856) the visions were there again, intact and less easy to manage than previously.

The first waves of alarm focused on the general competitive qualities of the Russian Empire, its emerging navy, the completely different society and the threat to links, within the British Empire. Later on more detailed military scenarios emerged. In 1868 Sir Henry Rawlinson, politician and soldier, invoked the vision of Russia as advancing militarily against India like an army investing a fortress.[11] He delineated several concentric circles in Russian strategy (including a line of 'observation', which had already been laid out, and a line of 'demonstration', which would soon be enacted) analogous to the more familiar small-scale practice of a siege. When the Russians dispatched a provocative military expedition to the Afghan border, the British invaded Afghanistan (1878–1882). Russian troops advanced no further, but fear of a Russian invasion of India persisted until the Anglo-Russian Entente was concluded in 1907.

On a global scale the image of lines encircling the world with points of support applied rather to the structure of the British Empire, and the image of a fortress applied to the Russian Empire. The British lifelines had long remained unchallenged by other sea-powers but they were vulnerable if a strong continental power – the fortress – should take control of the continental periphery and its points of support. This, of course, was Mackinder's vision expounded at a meeting of the Royal Geographical Society in 1904. The military strength of continental powers had increased comparatively in the preceding decades because of the development of railways. Rail transport enabled the deployment of troops to borders as far apart as the Pacific Ocean and the Mediterranean, much faster than sea transport would allow. Peripheral railways such as the Trans-Siberian Railway were also vulnerable. But a continental power ('the Heartland') could shape a radial transport structure which was much more resistant to disruption by military action from outside. The technical and financial impediments to developing a defensible infrastructure in Russia at the time were omitted from consideration in this futuristic lecture.

Mackinder's vision fitted in very well with mounting doubts about the defence of the British Empire, and its counterpart, the security of the home territory. In *The Battle of Dorking* the war was lost because the fleet was dispersed, fulfilling its duty to protect the empire in places far apart, responding to American threats to Canada, suppressing a mutiny in India, and so on. The clear message was that to save the empire and maintain national security at home, alliances were desirable. Mackinder's geopolitical model provided an intellectual script and geographical model for political action, but actual decision-making in foreign politics did not follow this straightforward opposition between land and sea power.

GEOPOLITICAL TRANSITIONS OF THE TWENTIETH CENTURY

The 1902 treaty between Britain and Japan actually preceded Mackinder's treatise and was the purest illustration of its principles: the alliance between two sea powers in the 'Outer Crescent'. As regards other power relations a terrible dilemma loomed ahead. First, since the rise of the German Reich two continental powers could now be associated with the concept of the Heartland: Russia, containing the Heartland, and Germany, able to assimilate it. Second, the strongest power in the continental periphery was a traditional enemy, France. The defeat of Russia in the Russo-Japanese War of 1904–1905 changed the image of Russian threat and provoked the Anglo-Russian Entente of 1907, an 'impossible' alliance. Earlier, in 1904, another 'impossible' agreement (though not in the Mackinder sense) had been concluded with France. Thus, in 1907 Britain had obviously chosen its enemy.

These events constitute what Peter Taylor has called a *geopolitical transition* separating two periods with distinct world orders and involving fundamental changes in popular images of the world.[12] The old nineteenth-century order was characterized by an almost unchallenged British hegemony and separation from European political affairs. The next period, with its Anglo-French-Russian entente, was described as a 'partial commitment' by the British themselves. It meant that the isolationist outlook was maintained even while the country was being irresistibly sucked into Europe, ultimately into the two most devastating wars in world history.

The next geopolitical transition occurred at the end of the Second World War and the subsequent world order became known as the Cold War. Although the focus is usually on the main antagonists, the United States and the Soviet Union, a special role has been attributed to Britain during the gestation period of the Cold War, between 1945 and 1947. It can be argued that the Cold War world order particularly fitted the British desire to continue its status as a world power within the new, precarious, conditions following the end of the war. Welcome as the victory was, it left a country with reduced powers and in economic crisis, as Winston Churchill, elaborating on frequently voiced British fears, had anticipated in 1937.

The principal concern of politicians like Foreign Secretary Ernest Bevin was to safeguard Britain's position as a power of the first rank in the postwar world. This position was severely threatened in several ways: the British economic collapse as a consequence of war efforts and damage; the head-strong actions of parts of the empire including Australia and New Zealand; the decisive role of the United States and the Soviet Union in achieving victory; and the emergence of the ultimate weapon, the atom bomb, in foreign hands. The 'Britain hypothesis' (Taylor) holds that the British engineered the bipolar structure of the world – afterwards known as Cold War – to defend their own global interests. There is no definite proof that American policy would not have evolved in the same direction without British instigation. But it can be agreed, on several grounds, that the United States was not immediately heading for this new antagonism at the termination of hostilities. First, Britain and the United States were not natural allies. The one represented the principles of colonialism and 'imperial preference' in trade relations, the other championed free trade and the dissolution of colonial empires. Second, the reverse applies to the relations between the United States and the Soviet Union: there is no profound geographical or economic reason for mutual fric-tion or antagonism, as American officials themselves declared immediately after the war. And, finally, the United States needed some time to become aware of its newly acquired greatness and to accept the leadership role it had refused on earlier occasions. The sum of all this was that the US world view inclined to universalism and the British view to the delimitation of spheres of influence: a recapitulation of the nineteenth-century distinction between Britain and the continental European countries, but with Britain now in a reversed role.

Indeed, there are unambiguous Cold War declarations *avant la lettre* by British politicians. In a well-known speech in Fulton (Missouri) on 5 March 1946, Churchill introduced the image of an 'iron curtain [which] has descended across the continent.' Bevin, about the same time commenting on Russian intentions with respect to Persia, evoked the model of Hitler's strategy of conquest: taking one position at a time. If Persia would fall, Turkey would be still less defensible. This was obviously an early formulation of the 'domino theory'.

What were these images supposed to accomplish? The emphasis on the Russian danger would help win over the American Congress to agreeing generous financial support for Britain and it would invite the collaboration of the American fleet in defending the British Empire. The empire now became a *cordon sanitaire* around the Soviet Union. As Peter Taylor notes: 'To obtain American support, Foreign Office officials were prepared to make a major sacrifice: they would welcome the Americans on equal terms in an area they had long considered their special reserve – the Middle East.'[13]

The evidence of early Cold Warriors is not in itself conclusive. One may just as easily find signs of an early Cold War mentality in the United States as in Britain: the pre-war tradition of the 'Red scare' in the United States, the popularity of geopolitics during the war, or the formulation of a 'domino principle' by America's ambassador to the Soviet Union, Averell Harriman, for example. As early as 1944 Harriman stated: 'If the policy is accepted that the Soviet Union has a right to penetrate her immediate neighbours for security, penetration of the next immediate neighbours becomes at a certain time equally logical.'[14] This is not the place to enter into a lengthy discussion on the origins of the Cold War. I shall therefore finish this rendition of its history by mentioning two facts that seem beyond doubt. The first is that the Cold War was a logical solution to British problems. The second is that British diplomats in the United States, immediately after the war, emphatically disapproved of the course of American foreign politics.[15]

Yet, one problem connected with early Cold War voices demands attention in our discussion. This is the question not of the geographical but of the social origin of Cold War ideas. Are geopolitical visions manipulated from above or are they derived from popular feelings? Taylor assumes that the popular images of the world evoked in the geopolitical transition to the Cold War result from the machinations of politicians and their expert advisers. According to Gallup polls in 1944, a 91 per cent majority of the British public favoured the Big Three 'co-operating together after the war.'[16] Would it seem, therefore, that the shift to a more antagonistic course of British foreign policy within just a few years can hardly be attributed to the public's vision?

Other commentators on the interaction between public feelings and political goals during the war are more cautious. Bell concludes that in spite of the widespread admiration for the Soviet Union in 1942 and early 1943 there were shifts and also increasing doubts about Soviet intentions after the war.[17] Although most people preferred co-operation with the Soviet Union after the war, only a small majority thought the great allies would indeed co-operate. Bell also concludes: 'despite the wartime conditions of censorship and propaganda, public opinion had an autonomous life of its own, and was sometimes fully capable of following its own path.'[18] Actually, the Ministry of Information could not prevent the extension of British admiration for Russian fighting power to the Soviet regime itself and even to the person of Stalin. Tensions between public opinion and policy occasionally arose, but this did not necessarily mean that policy-makers would change course! The counterpart of this argument is that congruence between foreign policy aims and public views does not necessarily prove that the public controls the decision-makers, or the reverse. In such threatening times as war, events will tell a rather uniform message because there is one overarching common interest: to win the war.

Nor would British diplomats in the United States in 1945–1946 have agreed with the manipulation view. Ambassador Lord Inverchapel wrote that anger over Soviet politics was first felt not by American government leaders but by the mass of ordinary people who 'derive their views on events abroad by a sort of intellectual osmosis from the headlines and their favourite wireless commentator'.[19] As Donnelly, a British consul, established (on the very day of Churchill's Fulton speech), the change in US foreign policy was:

'a remarkable example of the democratic process at work because the driving force has come not from the top but from below. Events and public opinion have forced the obviously uncertain and reluctant administration into affording to the world at least some measure of the leadership which the United States ought to be providing.'[20]

Such perceptions may, of course, be unconsciously moulded by the wish to produce solid arguments for disagreeing with official American policy. But the impression of surprise, and the apparent desire to explain American 'democratic' impulses, caution against dismissing these perceptions too rashly. Donnelly noticed that Stalin's speech of 9 February 1946 had made a great impact in the United States, although it had received little attention in Britain. He suggested that this was because 'Americans attach more importance to speeches and declarations than do the more realistic and empirical inhabitants of the United Kingdom' and also because Stalin had attacked capitalism, 'the nearest thing to a universal religion in the United States'.[21]

The most probable interpretation of these voices is that certain actions of the Soviet government in 1945, such as Stalin's anti-capitalist speech and the Soviet rejection of international (UN) arrangements and judgements (on the Iranian question), had, on each occasion, provoked sharp comments in the American press, but that the policy-makers were still wavering, hoping for signs of a more co-operative attitude. Nonetheless, we may assume that there was a public willingness in the United States to subject the actions of the Soviet Union to an absolute test and that the repeated failure of the Soviet Union to pass the test ultimately helped to produce a Cold War reaction under President Truman whose vigour surprised even the British.

For the British, the new Anglo-American partnership meant the continuance of the illusion of separation and difference with regard to Europe. Despite the formation of NATO, Britain did not move closer to its European neighbours (Ireland included). The focus of its defence interests remained the Atlantic and the Middle East.[22] The asymmetry of this relationship was never fully discussed. What the British saw as a new period of Anglo-American hegemony, was interpreted across the Atlantic rather as the special connection with *Europe*.[23] Britain was the tutor of America in its new world role – albeit certainly an appreciated one – rather than an equal partner. American obstruction of British and French plans for intervention in Egypt in 1956 was a clear sign of the unwillingness of the United States to become associated with European world-politics, whether on the British, French or, for that matter, the Soviet side. Crises like the Suez affair emphasized the ideological separation of Europe and the United States, but Britain was only reluctantly to draw the obvious conclusion that its fate had become inseparable from that of continental Europe. As long as the Cold War continued, there remained at least a pragmatic element in the Atlantic connection. The latest geopolitical transition, the events in Eastern Europe, even thwarted that ultimate argument.

THE IDEOLOGY OF SOVEREIGNTY

As stated earlier, losing one's identity in the current transnational system is a fear besetting not only the British. However, to a greater degree than their neighbours the British have to cope with 'post-hegemonic trauma' and the impact of a geopolitical transition. It is not surprising that novels picture the future annihilation of the British 'race', that political leaders emphatically praise the virtues of British sovereignty, and that others still stress the difference between a free and pragmatic Britain and a bureaucratic and ideological continent. The aim of this book is not to denounce such views, although it is difficult to resist the remark that the self-image of being free from ideology is just as likely to be biased as any system of more explicit beliefs about the world.

The British statesmen who established the main outlines of national policy shortly after the war had all been brought up, one could almost say brainwashed, with a world view that presented the British Empire as a natural and unchangeable fact. The impression of immutability was persuasive if one had learned history from books in which the important events in the evolution of the nation had all taken place before 1815.[24] The idea of competition between emerging industrial nations was completely lacking in this intellectual formation. Popular history and books for young people displayed the 'Kipling attitude': 'The Anglo-Saxon, resourceful, fearless and undaunted, makes the road, builds the bridge, cleans out the fever swamp and rescues the disordered "lesser breeds" from their petty hates and love of darkness.'[25] Britain was simply the most advanced nation, the once and future model for all other nations.

Those subjected to such attitudes and visions in their formative years had ample scope to learn otherwise when they were faced with the Great War of 1914–1918, and again with the rise of Germany in the interwar years. Yet, even the Second World War could not quite erase this attitude of British superiority. The wars may have shown Britain's diminished power, but history could be read in a different way. After all, Britain had turned out to be a victorious power, not because it was a warrior nation but rather because this outcome suited its role in the world; it was apparently its destiny to be superior. If the interwar years had taught anything, it was that Britain would have to play a much stronger and leading role in the world. That meant leading politically, not raising its economic performance. The challenge was to continue the steadfastness that had characterized its entry into the war, 'standing alone' against overwhelming odds. The passive attitude of the interwar years was not an indication of diminishing power but of diminished vigilance.

There were other inhibitions to facing up to the economic situation and recognizing the necessity of creating a competitive economy after the war. The idea that Britain had emerged victorious from the war, only to find itself in a weaker position than before, was simply unacceptable.[26] It seemed illogical and unfair that the hard-won gaining of military security in (that is, separated from) Europe should immediately be followed by a new effort to safeguard national prosperity by cultivating European connections and curtailing commitments with the empire. These inhibitions worked on a subconscious level. On the level of political discussion the situation appeared very straightforward. 'It is not ideology, but logic and commonsense, plus the brute facts of Britain's world-wide entanglements' that prevented Britian from forming a bloc with other West European states, declared Christopher Mayhew, Minister of State, in 1950. Seventeen years later he conceded that he had been wrong at that time and had not seen that to play a role in world affairs was already beyond Britain's real powers.[27]

Michael Blackwell concludes that British policy-makers did not make many really wrong decisions in the aftermath of the Second World War. Actually there were many reasonable arguments for the choices they made. A rapid rundown of overseas commitments could have created international chaos and would also have involved a selfish and cynical stance towards the poorer members of the Commonwealth. The mistake, Blackwell says, is not that policy-makers did not immediately make major adjustments but that they delayed recognition of the need for adjustments in the foreseeable future.[28] There were several real impediments to British integration in Europe immediately after the war. Even on the continent official support for federal futures was lacking. The military defence of Britain demanded priority for measures against attack by air, whereas continental defence strategies focused rather on land campaigns. The, at first, higher standard of living compared to that of most other parts of Europe did not seem to make membership of a European common market a very urgent matter. And finally, the status of Germany, still subject to Allied control, complicated the question of the implementation of a European federation. These facts did not encourage early attempts to attain a common Europe. At the same time, however, a fundamental discussion of the desirability of integration was evaded. An affective screen was lowered between Britain and continental matters, diminishing the relevance of all agenda items with respect to Europe, including those concerning long-term prospects. With pejoratives like Bevin's 'talking shop' for anything hinting at a European Community, serious proposals and analysis were held at bay.

At the end of the 1960s the inevitability of closer European co-operation and even membership of the European Community was more clearly seen, but affective inhibitions continue to play tricks on Britain's foreign policy strategy to this day. As Wallace observes, the British self-image as being pragmatic – for example, in considering *our* foreign policy versus Europe as based on a 'Europe of the facts' in contrast with *their* 'Europe of the phrases' – is completely at variance with the deeply symbolic nature of the so-called 'special relationship' between Britain and the United States. In fact, long-term trends are re-orienting the United States away from Europe.[29] Thatcher's emphasis on national sovereignty, invoking the principle of subsidiarity in Europe, was inconsistent in that it ignored the contrast between the commitment to regionalization of the European Commission and the British tendency to represent Scotland, Wales and England as an indivisible unity. The British reaction to Germany's new power – comparing it with Hitler's plans for Europe, or revealing itself in *Schadenfreude* at Germany's rising inflation rate after the unification – was equally emotional.[30] The impression emerges that British policy-makers still have difficulty internalizing European problems such as those involved in the expansion of EU relationships with the countries of Eastern Europe.

Are there no special British interests, then, requiring a specific British foreign policy? One of the interests defended by Britain's foreign policy-makers in recent decades has been free trade at the global level and the elimination of

Figure 2 'At last! Our dream is coming true.' Britain and continental dictators (1987). Cartoon of the British anti-EC movement (artist unknown)

protective arrangements within the European Community. This fits the model that compares Britain's relations with the closed, statist Continent with those of Hong Kong *vis-à-vis* communist-ruled China. Britain keeps the door of the European Market open for Japanese trade while benefitting from 50 per cent of Japanese investment in Europe. One might, however, observe that even this ideal of free trade on a global scale is unduly influenced by imperial reflexes. As Wallace notices, the United States and Japan have been negotiating an agenda including domestic regulation and patterns of public expenditure that goes far beyond the traditional agenda of international trade in its intrusiveness.[31] The game of international trade is changing and Britain will probably be the loser without collective arrangements like those of the EU.

The earmark of the new situation is not, what is most feared by the public, that it demands the sacrifice of national identity. Indeed, a particular national identity, a sense of location in the world and the feeling of sharing a national tradition, is a prerequisite for being able to contribute in a meaningful way to a wider community, on the national and the supra national scale.[32] Certain British traditions, like the system of political representation, international diplomacy and perhaps even pragmatism, may become the particular British contribution to Europe – if they have not already. Other legacies from the past, such as Britain's geopolitical vision, are sometimes more counterproductive and will undoubtedly have to change in the long run. As David Marquand states,

The British state was the child as well as the parent of empire. It was created for empire and lived by empire. Its iconography, its operational codes, the instinctive reflexes of its rulers and managers were stamped through and through with the presuppositions of empire. By the same token, the identity it claimed to embody and helped to create was, of necessity, imperial, oceanic, extra-European: it could not be anything else.[33]

In this sense Britain is one of the most illuminating examples of geography, or rather of a human geographical system, creating an obstinate geopolitical vision.

4

THE MARCH OF CIVILIZATION
DESTINY AND DOUBTS

(USA)

TWO EXPLOSIONS IN AMERICA

At 9:02 a.m. on 19 April 1995, a bomb destroyed the Alfred P. Murrah federal office building in Oklahoma City. The entire façade of the building was blown off, crushing 163 people, including some children in a day-care centre. It was the most violent terrorist attack ever committed within the national borders of the United States. Furthermore, it occurred at a location that seemed to deepen the experience of shock: 'The Heartland Bombing', said the headlines. That phrase revealed the general perception of terrorism as a danger that comes from the outside. New York – the frontier town between North America and the rest of the planet which should be allowed by the rest of the continent to 'float out to sea'[1] – would have been a much more likely place for the ghastly spectacle of a bomb attack (which it had indeed experienced some years earlier).

Initially all guesses as to the identity of the culprits tended towards foreign terrorists, especially Islamic fundamentalists. Soon, however, the collected evidence enabled the police to trace some very American suspects. Their crime might even be considered the ultimate consequence of a belief in American values: a belief in individual freedom, in the right to use violence against any person or force threatening that freedom. The particular outcome of the belief in this instance was embarrassing, because flirting with the idea of attacks on federal institutions was apparently flourishing at the time in circles in which anti-government rhetoric had the ring of common sense.

The main effect of this denouement was a deafening silence: no witch-hunt, no movement for moral reform (so often the preferred reaction in the public's acknowledgment of evils in the United States), not even a large federal research project on political attitudes was announced. How different from the aftermath of the *first* bomb attack in the United States, the Haymarket bomb!

On 4 May 1886, a peaceful demonstration for an eight-hour working day at the Haymarket in Chicago was drawing to an end when a group of police-men marched into the square and summoned those present to stop the meeting and disperse. In view of the previous restraint of the police, when the gathering had been much larger, this conduct clearly lacked legitimation. Completely unexpectedly, a bomb was thrown into the group of police officers, killing at least one. Seven policemen and an unknown number of civilians were fatally wounded; in total sixty-seven casualties were registered. But most of the injuries had been caused by bullets rather than by bomb fragments. After the attack the police started to fire indiscriminately into the crowd, wounding not only many innocent onlookers but also a number of their own kind.

The same spirit of blind revenge continued to dominate the search for suspects after the event. All the speakers and anarchist leaders were arrested, and many bystanders, even if they could clearly prove that they had not been involved in the assault or its preparation. Public opinion was deeply inflamed. As Paul Avrich states:

A fear of subversion seized the country, triggering a campaign of radical-baiting rarely ever surpassed. It was the first great American inquisition since the Salem witch-trials of the seventeenth century. For weeks and months the country remained in the grip of hysteria.[2]

That the majority of the anarchists were foreigners (that is, German immigrants) contributed to the general fear of a conspiracy to destroy the city. 'The "enemy forces" were not American, declared the *Chicago Times*, but "rag-tag and bob-tail cutthroats of Beelzebub from the Rhine, the Danube, the Vistula and the Elbe" . . . the "offscourings of Europe".[3]

The bomb-thrower was never traced; it may have been a policeman.[4] The leading anarchists were sentenced on the indictment that if they were not the makers of the bomb they were at the very least its originators. Five were condemned to death merely on the charge of 'exciting the people of this city to take the lives of other people'. If one's public writings could be interpreted as part of a conspiracy with people one has never met, then 'I can't see why we shouldn't be held responsible for any mischief whatsoever in the world,' remarked August Spies, one of the condemned men.[5]

Why was the social and official reaction so different in these two cases? Why no 'moral panic' or refurbished 'red scare' in 1995? A first suggestion is that the quality of justice, law enforcement and social knowledge has improved greatly during the twentieth century. No contemporary judge or jury would condemn a person merely for his or her ideas even if such ideas had motivated the crime of another person. But are we that sure? An affirmative answer seems to ignore that the US House of Representatives Un-American Activities Committee operated barely one generation earlier. Even in the 1950s people who appeared before the Security Hearing Board (to vet people in government services) had to answer questions like 'Have you ever indicated that you favoured a redistribution of wealth?' or 'At one time or two, you were a strong advocate of the United Nations. Are you still?'[6] Some critical observers might even wish to refer to more recent judicial derailments in the drive for 'political correctness'.

A further explanation of the different public reactions to the two acts of violence is that police investigations soon eliminated all speculation concerning conspiracy in the Oklahoma case. The suspects were found to have acted of their own accord and because of particular personal traumas. This conclusion was accepted in the press with a sigh of relief and with surprising docility. Not that the denial of a conspiracy was in itself unjustified. But the conclusion seemed to be all too welcome in the way it cut off searching questions into the ideological background of the offenders. The move into terrorism may have been an individual aberration but the deed symbolized a state of mind, shared by many others, that in many other countries would be considered a straightforward threat to national security. These 'new revolutionaries', as they were labelled in a review by Garry Wills, are against taxation, government regulation, public education, and state police forces, and they demand free ownership of guns.[7] They see state agencies as actors in a grand conspiracy aiming at the subjection and enslavement of free citizens.

One wonders whether the conclusion that the Oklahoma offenders acted on their own initiative would have been equally comforting if they had turned out to be Arabs. The opportunity to discern a 'foreign' element seems one of the crucial factors behind the Haymarket hysteria. Then as now, political leaders yielded to the temptation of managing domestic disorder by highlighting its foreign inspiration. Certain international relations scholars[8] even consider foreign policy-making as a way of expressing and constructing national identities with the aim of disciplining domestic society. This symbolic conception of foreign policy, as a 'boundary-producing activity' rather than as a pragmatic response to outside threats or opportunities, presupposes a distinct field of international action and credible antagonists. These conditions were not quite fulfilled in the United States for much of the nineteenth century. First, the history of colonization and territorial expansion did not attune Americans to the kind of multilateral internationalism prevailing in Europe. Second, the principal American image of otherness was derived from something very close to hand, the native American. Describing strikers and labour activists as 'savages', as people displaying the features of 'tribalism', was almost second nature to the elite press.[9] With the fresh experience of life-and-death struggle on the national territory there was little need to dramatize international threats. The experience of the Indian wars and the Civil War explain the belligerent language addressed to the strikers and the unabashed suggestions of exterminating the 'dangerous classes'.

FRONTIER LINES OF IDENTITY

American identity was of course constantly challenged by the inflow of different ethnic groups. But whatever the ethnic prejudices evoked by events like the Haymarket bombing, they never barred any group from civic participation. For some reason this participation in the construction of America melted 'individuals of all nations into a new race of men', as Crèvecoeur had already observed before the American Revolution.[10] Whether this melting process is actually completed or ever will be, one might assume that it required intense symbolism and the invocation of geopolitical visions to shape and reinforce American identity. Two 'frontier lines' symbolized this identity in the nineteenth century: the shielding against European interference, declared official foreign policy in what became known as the Monroe Doctrine (1823), and the much acclaimed '*frontier*' itself, a moving zone separating the civilized world from the wild, unsettled, Indian area. But these lines implied both non-territorial ideas and rather inarticulate geopolitical visions and realities. The Monroe Doctrine once more emphasized American independence and difference from Europe, while the frontier primarily implied the onward march of civilization in an almost abstract space. Both concepts fitted the United States' vision of itself as a manifestation of the westward march of civilization, a movement from the world of darkness into the light. The idea, labelled Manifest Destiny in 1845, could be extended, in that the Pacific and perhaps the Asian lands beyond were the natural destiny of America. However, only at the end of the nineteenth century did this vision of a westward pilgrimage evolve into the self-image of a unitary world power with foreign policy goals other than independence from Europe.[11]

At the time of the Haymarket bombing, the Union had become a territory of contiguous states; no fact better illustrates that by this time the frontier had become history than the publication of Frederick Jackson Turner's study *The Frontier in American History* (1893). Turner assumed a profound influence of the 'frontier experience' on the formation of American character and society. The frontier thesis holds that Metropolis (the mother country, and subsequently the American core area) becomes transformed and energized at the touch of an outside force. It evokes the image of a movement from an unexplored domain of abundant natural resources to a metropolitan core dominated by scarcity; of the purifying influence of virgin lands demanding to be organized from scratch without the deadwood of tradition; in brief, the thesis identifies the American elixir of youth. But the thesis also built a myth in ignoring that western modernization was generally an urban and industrial phenomenon, never a product of agricultural frontiers, and that one of the main mobilizing forces in American society was the steady inflow of cheap labour (immigrants).

The frontier idea was not merely an innocent conception of social rejuvenation by means of human endurance and hardship. It was explicitly connected with violence. Violence became celebrated in the myth itself and its image was diffused world-wide through popular media such as the western movie. This mythical tribute to *regeneration through violence* is somewhat rashly characterized by Richard Slotkin as an 'exceptional' feature of American culture. There are, however, examples of other nations who have likewise exaggerated the role of violence in their own history,[12] but the American myth is possibly exceptional in its ambiguous relation to the 'primitive', Indian world. The frontiersman does not dominate simply by his command of civilized technology but also by individual skills acquired in the wilderness, even learned from the Indians themselves, who are sometimes allowed to participate as good savages. This romantic vision that Americans, in contrast to Europeans, are able to renew their social vitality and culture by reverting to natural principles of individual survival, remains vivid today. John F. Kennedy took advantage of the idea in creating the Green Berets (1961), who were assumed to adopt guerilla tactics in Vietnam to defeat the

Reds. According to a journalistic impression each Green Beret had detailed practical knowledge ranging from the bow and arrow to the howitzer, and had studied the works of Mao Zedong and Che Guevara as the most up-to-date texts in his field.[13]

The awareness of the vanishing frontier, a nineteenth-century American variant of the 'end of history' theme, initially scared the middle class in view of the proletarian discontent in the large cities. How could such collective eruptions be diverted if the safety valve of cheap lands to the west was blocked? There was no answer. But the course of American history ultimately showed that something like a frontier experience could be very well transferred to other domains such as industrialization, acquisition of personal wealth, science, and foreign markets. Even the 'frontier country' of the ghetto could become a challenge or test of proving one's citizenship by escaping and making a 'typical' American career.

Whether Turner only helped to build an American myth or revealed something of an actual social mechanism, the nineteenth century ultimately produced a new and 'expansive' American vision of history. Americans, in the words of Frances FitzGerald, came to 'see history as a straight line with themselves standing at the cutting edge of it as representatives of all mankind.'[14] Together with the ideas of human dignity and freedom – rooted in America's primal role as a refuge from suppression – these values and images constituted the centrepiece of the American political myth.

BECOMING ENMESHED IN THE WORLD

The sharp public reactions to labour unrest, as in the Haymarket case, seem to betray fears about imminent dangers to the national way of life and its supposed geopolitical framework. The experience of both extreme violence in the Civil War and the special sense of the 'end of history' connected with the closing frontier, had to inspire either visions of doom, if no new national purpose could be imagined, or visions of new challenges in space and time. Americans chose the road of renewal and expansion. The Spanish-American War (1898) heralded a new period of assertive international action which, however, never evolved into the real territorial imperialism that had characterized the colonization of the American subcontinent itself. At the same time, the world of the mind offered another opening. The period between the Civil War and the First World War produced the 'largest single body of utopian writing in modern times'.[15] Between 1865 and 1917 at least 120 utopias appeared, of which Edward Bellamy's *Looking Backward: 2000–1887* (1890) was the most popular. This book marked the start of a great wave of similar publications, many of which were devoted to the salutary effects of government control on equality, health and wealth. The number of such publications dwindled after 1900 and the genre disappeared altogether after the First World War. The improvement in urban living conditions at the turn of the century and the unleashing of the 'dark forces' of communism and fascism in Europe made utopias appear naive and pushed dystopia forward as a more sophisticated alternative.

The First World War did not imply a significant break with the geopolitical relations of the United States that had been established at the end of the nineteenth century. Certainly, the war was an interference in European affairs, but the decision to enter that war had come late and was afterwards judged by many as unnecessary or at least as an historical exception. As Colonel Edward House, President Wilson's advisor, wrote at the time: '[a German victory would] change the course of civilization and make the United States a military nation.'[16] Entering the war to prevent militarization! How strongly classic American views still resounded in this maxim. What happened after the Second World War, however, was precisely the large-scale global military commitment which had earlier been renounced. The events of this war Europeanized the world by extending the balance-of-power game, formerly confined to the – in Jeffersonian terms – 'destructive' European powers, to a global scale with America as the most important new player. It meant the end of a tradition of 'isolationism', or rather unilateral American action in the world. By entering into an alliance with European states the United States was re-playing the development of Britain's international relations a century earlier.

It is remarkable that America, at the zenith of its national power, staged the most extensive 'evangelism of fear' in its history: the anti-communist witch-hunt of the 1950s. David Campbell considers the American fear of anarchy and impingement on its identity as a persistent factor in American history and a principal inspiration of its foreign policy. In this respect the Cold War is an 'important moment in the (re)production of American identity, that was not dependent upon (though clearly influenced by) the Soviet Union for its character' and 'any number of historical events might be considered as precursors to the Cold War.'[17] One might disagree with this point of view for various different reasons, but the main comment to be made here is that Campbell's explanation blurs the unique historical significance of America's postwar situation. In the First World War the United States had intervened to restore international equilibrium. In the Second World War the United States was *caught* by an unfolding war and attacked from an unexpected direction (Japan). After its victory, the United States could not withdraw as it had done previously. In order to prevent a repetition of the disastrous course of affairs after the Treaty of Versailles (1919) it wanted in the first place to take care of Germany and Japan and to keep in touch with those forces that favoured stabilization. Even with its newly won self-confidence, the United States still had no experience of international responsibilities. These could be seen as frightening because they involved interaction with powers like the Soviet Union who would, no doubt, be unimpressed by the American social system. America's presence in the world might easily assume the shape of a contest. In view of the American self-concept as 'standing at the cutting edge of history as representatives of all mankind', there was no need to worry about the outcome. However, America's performance might be subverted by individuals and groups who did not satisfy the criterion of American-ness. In this new international context the American values of freedom and superiority actually backfired and caused an extreme form of social control.

Figure 3 The difference between the hemispheres
(John T. McCutcheon, 1920)

The above account is not entirely inconsistent with Campbell's argument because it assumes the basic American fear of disunity, but at the same time it incorporates the special geopolitical situation after the Second World War. This situation, more precisely the presence of the Soviet Union, was not simply one of those external facts which happen to be suitable for addressing and shaping innate fears in society, it was an essential element in their production. The actual outcome of this was the historic continuance of a security discourse that laid remarkably little emphasis on external threats to state territory. Formerly this stance had been authorized by the geographical separation of the rest of the world, particularly Europe, and the overwhelming task of pacification of the American continent itself. After the Second World War the lines of territorial defence immediately expanded to other continents.

Figure 4 America and war-mad Europe (Carey Orr, 1933)

There has almost never been an extended period in American history during which feelings of (in)security focused on the national boundaries and narrowly identified with the contiguous national territory. The 'Other' was either far away or had already invaded America. Consequently the US boundaries could very well be seen as a transition zone and it is not surprising that initial reactions to a deed of terrorism in Oklahoma referred to the 'Heartland'.

But what about the difference between the aftermaths of the Oklahoma and Haymarket affairs? In Oklahoma the militants themselves thought that they were defending America, and the public did not show the kind of moral panic caused by the comparatively less violent Haymarket bombing. The absence today of recent domestic collective violence in the public memory is an obvious explanation of the moderate public reaction. Race riots are the closest corresponding experience of violence in the postwar period but they tend to be interpreted as caused by a condition of diminished responsibility in the offenders rather than as political violence and their magnitude does not compare at all with the American Civil War. Another explanation, already mentioned, is the impossibility of relating the Oklahoma suspects to some kind of international threat like communism or Islamic fundamentalism. But why was that impossible? The ethnic identity of perpetrators of 'dangerous' activities has never been an obstacle to 'discovering' un-American inspiration. There is only one answer: the nature of the deed goes too much to the heart of doubts besetting the public itself.

Since the disintegration of the Soviet Union and the communist bloc (1989–1991) Americans are, for the first time in their history, faced with a world in which no significant territorial line can be traced between Self and Other. There was never time for the only real dividing line, the federal border, to settle in American minds and hearts. Between the two world wars the United States was hardly a political participant in the world system, and after the Second World War the security lines followed a forward defence system spanning the entire globe. It is understandable that the collapse of the post-war geopolitical order should throw Americans back upon the family or the purity of an ideological community. By infringing the autonomy of ideological communities, as in the Waco tragedy of 1993 (when Federal Bureau of Investigation agents set fire to the compound of the Branch Davidian sect in Waco, Texas), the state has become a threat to the community instead of a source of security. This is what the perpetrators of the Oklahoma bombing enacted.

THE SHATTERED MYTH?

Political commentators nowadays agree about the declining faith in 'the American myth'.[18] But they do not see the militias, or other adherents of (inter)national conspiracy theories, as the last vestige of the myth. They report a waning of the common vision of the world after the fall of communism, and a social fragmentation in which the sub-myths of various sub-communities displace the American political myth. The new revolutionaries cannot provide any positive alternative to these problems, in fact they open the door to further fragmentation. Some visions of the world in these circles even sound strangely un-American: the vision of a Zionist–capitalist conspiracy to defeat America, for example. The radicals of the Right, however, have a point which derives its strength from an assimilation of lingering Left and Right anti-government resentment: the commitment to war of successive American governments has frequently involved the withholding of information from citizens, or even downright deceit. Another tenable argument that fits this viewpoint is that the postwar period can be seen as an aberration within an overall American tradition of (political) non-commitment on the global scale.[19]

America's new multicultural experience has produced a complete reversal of the classic frontier idea by emphasizing (sub)group identity and transforming the *challenge* of borders and frontiers into the idea of *violating* (some)one's identity. In a kind of outpouring of anger, Robert Hughes has portrayed the current American political scene as packed with self-centred groups, homo-sexuals, pro-life activists, Christian fundamentalists and ethnic groups who are bitterly complaining about wrongs suffered at the hands of unconcerned others.[20] This sense of victimization, of being raped (particularly by 'dead white-European males'), may ultimately justify violence, and possibly set off a prolonged civil crisis.

The current situation may be seen as just one of those transitory stages in the historic oscillation between introversion and extroversion in American politics.[21] The political focus may currently be shifting to domestic problems and this may change something in the public's vision of the world. No theory, however, predicts what these swings of the pendulum may imply for the contents of visions.

The idea of cycles in history provides only one possible interpretation of America's present predicament. Another perspective is to consider the postwar anti-communist crusade as an abnormality in a country which, as revealed by its history, is deeply committed to isolationism and individualism. But why should we have to choose between either regular oscillations or a primordial attitude to the world around?[22] Is it not more natural to expect that there should be events in the history of a nation which are so deeply disturb-ing or exciting that they define visions of the world for a long time? Such visions may be interrupted by other events, or possibly replaced, but not in a regular way. One has to wait until the combination of an external event (or crisis) and the arousal of a value profoundly rooted in national history and geography brings about the new and outspoken ideas concerning the world order that were still so poignantly absent in 1990 in America, despite George Bush's rhetoric.

Whereas Bush had started his presidency with the invocation of con-servative values like faith, sacrifice, moral principles and the identification of drugs as a foremost threat to the nation, Clinton's leading card was *renewal*. His inaugural address (1993) teemed with words suggesting a new start: renewal, reborn, rebuild, reinvent [America], Spring, change, revitalize, revo-lution, experimentation. These words seem to address primarily domestic problems. As for the world outside, it was becoming more free but also less stable. In line with Wilsonian tradition, Clinton finds the greatest strength of America in 'the power of our ideas', America as model for other nations rather than as a patrolman.

Both post-Cold War presidents drew from American tradition to define a new national purpose and to make some sense of the world order. The one fell back on moral and religious principles, the other on the frontier myth of endless youth. Neither attempt found a lasting anchorage in the new world. It is tempting to agree with what Henry Kissinger is saying in one of his recent writings: 'America is presently going through what I believe to be only the beginning of a national debate.'[23] Doubts and uncertainties are not new in American history, but whereas earlier generations could draw fresh energy from physical challenges and external destinies, the current one has to draw everything from within itself.

THE LAST FRONTIER

(USA)

DISCONTINUITIES IN AMERICAN GLOBAL PERCEPTION

One could easily be tempted to believe that America today is more complicated, that the country speaks less with one voice, than in the past; that contemporary Americans have lost the clear sense of purpose that guided earlier generations. However, contradiction and paradox have never been absent from the American character, as Arthur Schlesinger Jr has said. There always was a 'schism in the American soul between a commitment to experiment and a susceptibility to dogma', between realism and the belief in a special mission.[1] On the one hand, the old Puritan vision of America as a blessed refuge from the evils of the world (particularly those in Europe) and as a beacon of justice and freedom has become one of the most generally acknowledged features of the American self-image. On the other hand, the Founding Fathers also reminded their fellow countrymen of the risky nature of statecraft and the impossibility of forcing international realities into the mould of domestic ideals. The future was uncertain, and their confidence in the future was based on the supposed geographic and demographic advantages of America, rather than on divine intercession.[2]

The two points of view are not completely incompatible because – at least in the Calvinist approach – even God's guidance does not exempt man from the requirements of precaution and effort. God is testing his followers! Yet, in practical politics it may be difficult to combine realistic and missionary attitudes. If a definite choice between two valued principles cannot be made, a possible outcome is to alternate between both aims. The impulse for a policy reversal originates in the disappointments which result either from external resistance or from limitations inherent in a particular policy mode itself. In the international arena, the disappointments emanating from a strongly missionary attitude clearly accumulated in the Vietnam imbroglio.

The effect of America's victory in the Second World War was a perception of limitless choice, of 'having the ability to participate or to withdraw from international engagements at will'.[3] American interference seemed sufficient to determine the outcome of any conflict or problem in the world. Vietnam for the first time really shattered this belief. The typical reaction to such experiences is withdrawal (isolationism) or realism. Kissinger's geopolitics, a new realism, developed around 1970 during the withdrawal from the Vietnam entanglement.

Contrary disappointments arise from realism's need to act swiftly, the impossibility of raising full national support for subtle political balancing acts, and the unfounded assumption that other players on the international scene make similar realistic estimates. The apparently 'blind' eruption of domestic sentiments into international relations, never more terrifyingly staged than in the American embassy hostage crisis in Iran in 1979, is difficult for the realist (or geo-)politician to process.[4] The strongly assertive international style of president Ronald Reagan in the first half of the following decade is understandable as a logical reaction to the Iran crisis.

Undoubtedly practical foreign policy will always have to include elements of both realism and idealism, of reactive and pro-active politics, but even if some measure of cyclic alternation is accepted, does this also imply recurrent change between different geopolitical visions? An affirmative answer seems premature. There is no reason why a more introvert or realistic stage should not sustain the basic picture of good and evil forces evoked at a time of political crusading (and become obsolete in a succeeding stage). Public visions are more sluggish than practical politics and often more belligerent too. US foreign policy after Vietnam neither eliminated the negative image of communism as a dangerous political force, nor did it substantially increase understanding of nationalistic impulses elsewhere in the world, as the US's tardy recognition of the forces for local autonomy in the former Soviet Union demonstrated.[5] Changes in practical politics, however, could be unmistakably established around 1970: a diminishing willingness to submit to heavy sacrifices for the sake of other nations, the rapprochement with China, and a sharp decrease in the level of conflict with the Soviet Union.[6]

The Nixon–Kissinger strategy involved one or two elements that may count as a change of geopolitical vision. Most important was the dropping of the image of communism as a monolithic force. The other element was the recognition of the relative decline or limited power of America itself, and the ensuing caution in using power more selectively by taking advantage of (regional) balances of power. However, according to G.R. Sloan, these changing policies did not disrupt the basic *territorial* view of containment of Soviet communism that had dominated American foreign policy since the Second World War. Sloan even sees a continuity between all postwar American foreign politics until the 1980s and Mackinder's early nineteenth-century doctrine of the importance of Eurasia in the world balance of power. The message of both was that a margin of political and military superiority in the lands of the Eurasian rim was required to prevent the Soviet Union from becoming a dominant world power.[7]

After the Second World War, civil war in Greece, and the fear that for this country to succumb to Soviet influence would put overpowering pressure on Turkey, were responsible for the start of the well-known policy of containment, with the declaration of the Truman Doctrine in 1947. This vision pictured the countries of the Eurasian rim[8] surrounding the Soviet Union as 'points' that had to be connected by a 'line' to secure their ability to deter and withstand Soviet pressure. Dean Acheson, the intellectual author of the Truman doctrine, had used the metaphor of the 'rotten apple', but the argument became known some years later as 'domino theory'. The 'principle of the falling domino' was popularized by President Eisenhower in discussing the defensive chain of South-east Asian countries: 'You have a row of dominoes set up, you knock over the first one, and what will happen to the last one is the certainty that it will go very quickly.'[9] Few authors in the field of international relations have noticed that only children play dominoes this way.

The postwar policy of containment broke with the traditional American conviction that geographic separation from Europe was sufficient to warrant national security. Whereas the Second World War and the Truman Doctrine that followed produced a clear discontinuity, the effect of the next major experience of war, in Vietnam, was much more confusing. Its immediate and most painful result was the breaking of the postwar American consensus on foreign policy.

THE SIGNIFICANCE OF VIETNAM:
PRESIDENTS AND SECRETARIES

Wars and revolutions burden collective memory, and control national learning processes about the world, more than almost any other event, remarked Robert Jervis.[10] In view of its number of casualties, one would expect the American Civil War of 1861–1865 to be the most significant experience since the American Revolution. Its influence can probably hardly be overestimated, although effects on today's society and national character are difficult to establish unequivocally. An idea with some relevance to this discussion is the suggestion that the American Civil War was a total war, aiming at the complete incorporation or subjection of the adversary, and that consequently the experience would evoke the vision of 'annihilation' in each new violent conflict in which the United States became involved.[11] This would explain the use of the 'body count' instead of measures of territorial advance, and the practice of attacks on civilian instead of clearly military targets, during the Vietnam War. However, there is little evidence to substantiate the claim of a bias towards extermination in American practice of waging war in general.

Table 1 War trauma in the United States
(number of battle deaths)

Civil War (1861–1865)	618,000
First World War (1917–1918)	116,000
Second World War (1941–1945)	292,000
Vietnam War (1961–1973)	58,000

The influence of the Second World War or the Vietnam War on current political thinking and decision-making is easier to establish than that of the

Civil War. The significance of Vietnam does not depend on the casualty figure, but rather on the human toll in relation to the ambiguous purpose of the war. The war provoked an internal struggle about American identity and purpose without precedent in the one hundred years since the Civil War.[12]

During the first two decades after the Second World War the 'lessons of the 1930s' provided a tight model for American relations with the Soviet Union. During the Cold War the message of these events was that to appease totalitarian regimes (as with the Munich agreement with Hitler in 1938) was fundamentally wrong because it would ultimately result in war all the same – probably in an even more devastating form. The involvement in 'Vietnam' was initially defended from the same uncompromising stance. Yet a small group of war protesters learned a different lesson from the Second World War: that military tactics in Vietnam produced the very type of war-crimes revealed and condemned at the Nuremberg and Tokyo trials.

Richard Nixon (President 1969–73) was an old hand at Cold War politics[13] but it is the irony of history that the burden of disengaging from the Cold War's manifestation in Vietnam became the fate of his presidency. To stop the hopeless involvement in Vietnam, Nixon could not invoke the lessons of the 1930s. Thus it became his challenge to demonstrate the important ways in which the world had changed since the Second World War.[14] These included the growth in offensive capability of the Soviet Union, the narrowing of the nuclear gap between the Soviet Union and the United States, the emergence of new, assertive nations, and the dissension between the communist adversaries. The presidential narrative acknowledged the development of a more complex world, but in a contorted way this message also seemed to promise new scope for control and imaginative tactics. Indeed, a new note was sounding in foreign policy: Kissinger's revitalized geopolitics.

Geopolitics was described, by Kissinger himself, as an alternative to three dominant traditions in American foreign policy: the idealistic, the pragmatic, and the legalistic traditions. The geopolitical approach would be new in paying attention to 'the requirement of equilibrium'. 'Nixon and I wanted to found American foreign policy on a sober perception of national interest, rather than on fluctuating emotions that in the past had led us to excesses of both intervention and abdication.'[15] Although the word 'geopolitics' had indeed been avoided by postwar politicians and presidents, this manifesto of Kissinger masks the fact that preceding foreign policy traditions were also established on a firm geopolitical foundation in that they associated national security with territorial strategies and implicitly assumed the strategic importance of the Eurasian rim.

Kissinger's 'equilibrium' concept naturalized the existing international state system with its multiple players and stakes, while dropping any fixed geographical meaning. The preceding image of two bipolar powers with clearly demarcated spheres of dominance was refined and gave way before the existence of diverse regions with their own geopolitical niches and consequent legitimacies.[16] The morally despicable policy of defeating Iraq but at the same time keeping it strong enough to resist enemies like Iran (thus condoning Saddam Hussein's gross human rights violations) is a later variant, although military operations like Desert Storm did not belong to Kissinger's favoured repertoire.[17] In the wake of the Vietnam débâcle the promise of the Nixon–Kissinger strategy was that modest involvement could obtain maximum effect, at least if the US response was fast and candid. At an early stage, small but decisive efforts could take advantage of the natural propensity of an inter-state system to search for equilibrium. Pulling China out of isolation would compel the Soviet Union to pay more attention to its borders with China, which would subsequently alleviate the pressure on the West and on NATO. This approach made previous global anti-communist containment policy appear, as Melanson put it, 'lumbering and unimaginative'.[18]

The nature and laws of the global equilibrium, however, were never expounded by the protagonists of this strategy. A content analysis of presidential speeches reveals that of those of all postwar presidents, from Truman to Reagan, Nixon's speeches contained the smallest number of references to foreign places.[19] Nixon, at least, did not like to talk about 'balance' and 'equilibrium'. Much to the annoyance of Kissinger he preferred to win public acceptance by evoking the prospect of détente and by repeating phrases about a coming 'generation of peace'. On the one hand, the new geopolitics was actually the absence of a mission or aim in foreign affairs. On the other hand, the subtlety of the ideal of geopolitical balancing and the swift foreign policy reactions required, were difficult to sell to both the public and Congress. Therefore, Sloan may be right in claiming – although the claim is hardly underpinned with textual evidence – that the Heartland–Rimland vision persisted even during the reign of the new geopolitics.[20]

The administration of Jimmy Carter (1977–80) presented more audacious breaks with postwar tradition by explicitly criticizing containment policy and identifying it as one of the causes of the Vietnam débâcle. The Third World was Carter's new pampered child and the advancement of human rights in the world the United States' new mission. Notwithstanding its roots in the American tradition and its continuation of the multi-polar world vision, Carter's policy was not to survive the humiliation of the United States in Iran and the strategic shift caused by the Soviet presence in Afghanistan. Both containment policy and geopolitics, equally despised by Carter, seemed to be cruelly proved right at the very moment they were confidently being disposed of in the dustbin of history. The lessons of the 1930s had to be revived once more. Nazi Germany and its Olympics became the inspiration behind the boycott of the 1980 Olympics in Moscow.

Such historical lessons constituted a perfect springboard for Ronald Reagan's foreign policy. Basically resuming the post-Vietnam faith in selective and restricted interventions in other countries (including the inevitable collisions with Congress that had marked the Nixon era), Reagan added the suggestion of a massive attack on the Evil Empire. This idea implied more than mere containment. It aimed rolling back Soviet power, an aim intimated in the early postwar period by certain politicians but never implemented for fear of triggering a Third World War. Vietnam fitted this assertive model badly, all the more so since Reagan implied that a victory for democracy over totalitarianism did not require heavy sacrifices and was within reach. 'In general,

Reagan and his advisers tried to say as little as possible about the Vietnam War, usually lumping it with Watergate as an event that had produced national disillusionment and self-doubt,' concludes Richard Melanson.[21] Actually, Reagan and his circle knew very well that the ghost of Vietnam was not exorcised in the public mind. Huge opposition to the sending of troops to Central America testified to the disagreement between public opinion and practical policy aims. Reagan's spirit of optimism and his determination to efface a decade of 'bad news from abroad' was welcomed by the public, but the fear of military adventures going wrong (the 'Vietnam syndrome') had cut deeply.

The Vietnam syndrome is one of the most convincing illustrations of the fact that public opinion cannot be manipulated at will by presidents or politicians. Noam Chomsky defiantly redefined this derogatory concept as the 'technical term for the failure of the American indoctrination system'.[22] Notwithstanding Chomsky's positive reading, the influence of the Vietnam syndrome appears to extend only to direct military involvement in other countries. It does not necessarily imply a changed vision of the security dangers faced by the United States or of the danger of communism. There seems little reason to assume that the American public in the mid-1980s did not share the view of foreign policy specialists like those participating in the 'National Bipartisan Commission on Central America'. This group, chaired by Henry Kissinger, identified external forces influencing 'the crisis in Central America', such as: international terrorism, imported revolutionary forces, the ambitions of the Soviet Union and the example and engagement of Marxist Cuba.[23] That the geopolitical approach had not lost much of its vigour is demonstrated by the following passage:

> In short, the crisis in Central America is of large and acute concern to the United States because Central America is our near neighbor and a strategic crossroads of global significance; because Cuba and the Soviet Union are investing heavily in efforts to expand their footholds there, so as to carry out designs for the hemisphere distinctly hostile to the US interests; and because the people of Central America are sorely beset and urgently need our help.[24]

Lars Schoultz has summarized the tacit foreign policy beliefs behind such statements in four concise statements.[25] First, 'Hegemony is desirable'. Hegemony in Schoultz's vocabulary is crude power politics, such as covert action to subvert the election of a presidential candidate in a South American country. This kind of policy has often antagonized Latin American nations more than it has enhanced American influence. The general picture was an increase in American destructive power coupled with a proportionate decrease in influence. Second, 'US loss = Soviet gain'. To policy-makers who conceived of the world in bipolar terms even nonalignment was damaging to American interests. They could not see that even if a Latin American country was to call itself 'socialist', it would have to gain more by economic co-operation and good relations in an all-American framework than by isolation or joining the communist bloc. Third, 'Control of territory = Control of people'. The conception of the American continent as 'territory' only made good sense in the nineteenth century, when it could be used by European powers to threaten US security (Monroe Doctrine). Today Latin American nations consider this concept as an infringement of their territorial sovereignty. They do not accept the idea that their freedom to choose a political system or to trade with any other country should be limited by their location in the 'backyard' of the United States. Fourth, 'Supporting existing governments = Protecting US security'. By its support of loyal dictators, sometimes to the bitter end, US politics lost its legitimacy in the eyes of many Latin Americans. All these assumptions contributed to the deepening of the Vietnam entanglement but they still pervaded thinking about Central America two decades later.

Even if a direct military threat to the US territory could be ruled out – and Kissinger's Bipartisan Commission more or less conceded that revolutionary changes in Latin America were no real threat as such – the United States had to interfere to keep up its credibility. 'The triumph of hostile forces in what the Soviets call the "strategic rear" of the United States would be read as a sign of US impotence,' the Commission stated.[26] The first interest of the United States in Central America, according to the Commission, was 'to preserve the moral authority of the United States'. This is explained as conduct that shows the world that the United States does what is right *because* it is right. Translated into a more mundane language: the United States cannot waver in problems that others may perceive as a matter of principle, even if a retreat would be advisable for pragmatic reasons.

This approach turns the world into an American moral landscape. Countries enter the scene merely as testing grounds to prove American authority. But the conditions of the test (like Schoultz's four foreign policy statements), even the interpretation of the situation as a test, are largely a product of American imagination. During Reagan's presidency this enticed him into selecting places that would assure a victorious outcome to the power contest, rather than provide a response to high-priority threats to the Western world. Reagan was lucky, his successors had to face more serious threats.

THE SIGNIFICANCE OF VIETNAM: OTHER VOICES

One of the reasons for the Vietnam failure was the profound communication gap between two completely different worlds, two 'states of mind' as the American journalist Frances FitzGerald put it.[27] This gap, which never narrowed, also explains why the Vietnam experience hardly affected the American perception – official and public – of the world, and of the Third World in particular.

FitzGerald considers the American intellectual landscape largely as an inheritance from Europe, and that of the Vietnamese as derived from Chinese culture, although 'in their own independent development the two nations have in many respects moved even further apart from each other.' America differs from Europe because the urge to escape and to expand was never contradicted by physical boundaries or strong traditions. The consolidation of the national territory at the end of the nineteenth century only shifted the ideology of expansion into new areas such as industrial innovation, new foreign markets, and political dominance over other parts of the world. The national gaze fixes on the future and on shifting frontiers. An American will take you to the roof of a building and describe the view in terms of all his plans for the future while ignoring most of the present.[28]

For the traditional villager in Vietnam, place and earth had a quite different, existential meaning. To live somewhere was not a choice. Rural life did not encompass plural institutions addressing the native in separate roles, nor was deliberate change valued. Village and land were linked with ancestors and with personal identity. One could not exchange one's village with another without losing one's identity. Everything had a preordained place; family, village and state were intertwined in a fixed way.[29] The policy of removing people from their home villages in order to create space for military manoeuvres against communist infiltration violated the basic tenets of Vietnamese culture in two respects. First, it did not take account of the fact that the struggle was not a battle between two fixed groups of people, and, second, it knocked away the existential mainstay of people, making them more, not less, liable to the lure of ideologies, including communism. FitzGerald suggests that the Vietnamese villager could easily shift from one type of ideology to another if this would logically integrate his life in a wider and higher world with leaders presenting the right spirit and character. Vietnamese history had known revolutionary changes before.

The first lesson to learn from Vietnam would have been to drop the view of communism as a spatially progressive force, as a discrete and evil army impinging upon the freedom of resistant people. A related lesson would have been to recognize in communism an unpretentious way of solving the problem of the state in newly independent countries, not a declaration of hostility to the United States or 'the West'. However, in the aftermath of the Vietnam War, Americans were not particularly interested in what had motivated the Vietnamese in either fighting so inexhaustibly or yielding so easily to communist pressure. As far as Vietnam was represented in literature or films, it rather resembled rural family life in the United States itself.[30]

Did the experience of the Vietnam War not change the public vision of the world at all? Public opinion surveys certainly reveal a substantial impact, although poll data are usually inferior when it comes to establishing the (subconscious) myths and narratives which we apply to the world. Changes in public support for the Vietnam War and opinions on the mistakes of sending US troops to fight in Vietnam are well documented for the period from 1964 to 1973.[31] The figures reveal a steady decline in support for the Vietnam involvement, with 1967 as the turning-point when the number of opponents surpassed the number of supporters, a development that was never reversed. Also relevant is the opinion on the United State's commitment to security problems in the rest of the world. The Vietnam episode brought a sharp decline in the public's willingness to use force abroad. But it did not result in permanent isolationism. The percentage of people favouring an active role for the

United States in the world, including the use of military power to protect Europe against Soviet invasion, rose again after 1976.[32] The main impact of the Vietnam War on public opinion was to end the postwar consensus on America's role in the world. The public became divided not only over whether the United States ought to be involved in world affairs, but also over how it should react to external events. Wittkopf found two faces of internationalism in his study on public opinion changes, co-operative and militant.[33]

After the withdrawal in 1973, Vietnam was completely blanked out of public discussion and the TV screen, at least in documentary programmes. In one realm, however, Vietnam continued and even reinforced its presence: the world of the imagination. When political reassessments failed to appear, novels and films helped the American citizen to cope with a national trauma. The impact of the Vietnam War in the world of fiction was an infatuation with the psychical wounds of returned veterans, and an allusion to the special powers acquired by the Vietnam experience: a weird imperturbability or physical strength in individuals.

The war was not idealized as a national project; novelists and screenplay writers searched for a new meaning at the individual level. They tried to show how this 'frontier-experience', submerged in a terrible general confusion, would help to transcend common human limitations. Francis Ford Coppola's *Apocalypse Now* is a prototype in several respects. In the opening sequence of the film we see the hero as a heap of human misery in his Saigon hotel room. But very soon we are informed about the deeper meaning of this existence: he is 'waiting for a mission'. The mission turns out to be the killing of another American, a genius who has transcended the Vietnam experience in his own way, something deemed harmful by the military intelligence staff. As soon as the mission sets off, all symptoms of despair leave the hero and we see a balanced, rationally calculating individual. However, the artistry of script-writer and director have guarded against the tendency to create a Superman. In this case the human material stays believable.

In other plots more extraordinary individuals (Magnum, Rambo, Crockett in *Miami Vice*) fulfil their missions either in Vietnam, searching for lost comrades and veterans, or in America itself, purifying a degenerated American society. A third type of fiction merely paints a sentimental picture of disturbed family life and efforts to repair the damage. Vietnam itself stayed an abstraction, a non-place, aptly expressed in GI language by using the term 'world' for what the soldiers left behind when they came to Vietnam.[34] '"Getting used to" moving through the perils of time without the assurance of luck, without the conviction of a special grace conferred by a special geography, is precisely the function of the literary and cinematic narratives which American artists have produced in response to Vietnam.'[35] In these words John Hellmann sums up the message of the Vietnam fiction. Obviously we are witnessing the shattering of the American myth, or part of it, but also its reconstitution. The result is a confrontation of America with itself, a protracted soul-searching about American mistakes, and ultimately a tendency to lose oneself in tales of distant worlds and times and heroic battles (as in Lucas's *Star Wars* trilogy).

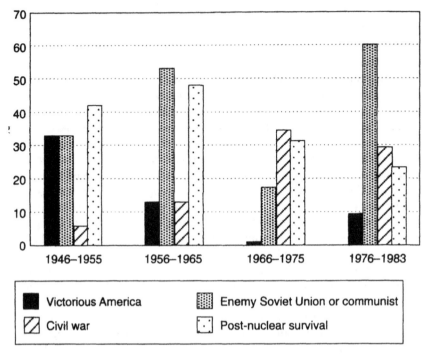

Figure 5 America in future wars: themes of 108 future war novels (percentages) written between 1946 and 1983

Whereas Vietnam fiction tells us little about changing visions of the world, indeed very little about Vietnam itself, the genre of 'future war' novel seems to offer better prospects for exploring changing perceptions of the national position in the world. The bibliography compiled by Newman and Unsworth (1984) provides a workable file. In this collection we find summaries of 191 novels written between 1946 and 1983 and published in English.[36] If we restrict our analysis to the novels focusing on threats to the territorial security of the United States, 108 titles remain. I have selected four themes by virtue of both the occurrence of sufficient variation across the different novels and their relevance for this discussion. The first theme is called 'victorious America', the outcome of the war in terms of a US victory as opposed to all other outcomes (not necessarily defeats). 'Victorious America' appraises the percentage of novels ending in the elimination of the enemy, although the term 'happy ending' does not always apply. Another obvious variable ('enemy Soviet Union or communist') establishes the nature of the enemy: is the enemy identified as the Soviet Union or communist troops, or is some other power involved? The third variable is less obvious. A substantial number of novels depict a state of civil war without relation to an extra-territorial threat ('civil war'). The fourth variable ('post-nuclear survival') establishes whether the tale is about human groups coping with the results of nuclear destruction, in some cases an event which happened in a very distant past, or about something else. Obviously the variables are not logically exclusive: America may be both victorious and have the Soviet Union as enemy.

Post-nuclear survival (or dying) is one of the favourite themes throughout the entire period of almost four decades, although a decline in interest is unmistakable from the 1960s on. These novels depict a disorganized world with wandering human groups and mutants killing each other or trying, like Robinson Crusoes of the future, to build a new, sometimes even idyllic, society. Many of the novels excluded from the analysis, because the scene could not be identified with the territory of the United States, would also qualify as post-nuclear. They are set elsewhere in the world or in a place without recognizable topography. As a matter of course in these novels, cities with a clear (national) identity have been destroyed after a worldwide nuclear war and national borders have lost their meaning.

One of the most striking trends is the declining faith in American power and the generally low incidence of the 'victorious America' theme. The interest in awkward outcomes of a possible war may be explained as a traditional feature of the genre.[37] The variation through the years, however, tells another story. The original European future war novels depicted war as a strategic game. The plots of failing military strategy and resources were aimed at drawing attention to blind spots in the national security vision and inciting politicians and the military to raise the national performance. This was the reality of late nineteenth-century Europe and it may also have been the mood in America for a short time after the Second World War. However, international developments soon revealed that the world had changed in significant ways.

Even Americans could not credibly resurrect on a world scale the conditions which governed the European balance of power at the height of its importance. The rise of nuclear technology had changed the world too much, certainly after the Russians acquired their A-bomb in 1950. Vietnam did the rest. The decrease in the number of war novels predicting an American victory had already begun before the Vietnam decade (1966–1975), when an American war victory became completely unthinkable. But this time the gloomy visions did not aim to provoke. The tales of disaster were not meant to drive home a political message on national defence, or the cleverness of a potential adversary. A key to understanding the concerns of many authors may be found in the theme 'civil war', the theme that mirrors that of 'victorious America' in Figure 5. An increasing number of novels depict war as a possible outcome of domestic struggles between cities, regions or races. Race riots and anti-Vietnam protest movements during the 1960s left their imprint on this literature. Yet, a similar shift of interest towards internal forces of destruction and domestic threats to democracy had occurred in the European future war novel a few decades earlier, between the two world wars. What does the American shift mean? The end of the genre? Or the awareness that the state as a viable political unit is nearing its end? In any case it implies the belief that war cannot be won without paying a disturbing price in terms of destruction, internal instability or loss of individual freedom. Indeed, the new conditions of strength – destructive power or social engineering – prevent the winning of a war in the traditional sense of saving democracy and civil society; they have become counterproductive.

The moderate increase of the 'victorious America' plot in the period 1976–1983 suggests a possible relationship with the prominence of the enemy. As long as the enemy (in this case the Soviet Union) is a recognizable power, the road to victory becomes imaginable. This implies that, in the American image, the enemy must be sufficiently similar to the United States, for example in geopolitical aims and military technology, to be a credible antagonist. All other enemies or hostile forces (Vietnam) are treacherous and unconquerable. Alternative enemies are sometimes made up of China, Russia and China, or some vague Eurasian alliance, but novels based on the Soviet Union as enemy far outnumber those with other (earthly) adversaries. The Vietnam War and the *détente* in its wake repressed the awareness of a Soviet threat to national security, but the invasion of Afghanistan and the first years of Reagan's presidency entirely rehabilitated this enemy in fiction. This also shows how susceptible fiction writers are to the political mood in the country. In view of this development nothing could have dealt a more terrible blow to the American world vision than the dissolution of the Soviet Union in the 1990s. As Clarke puts it in the second edition of his *Voices Prophesying War* (1992), 'For the first time in the course of this fiction, there are reports of vanishing plots, shrinking markets, and authors in search of new material.'[38]

POST-COLD WAR AMBIGUITY

The idea of a mission in the world has continuously characterized American foreign policy in the postwar era and the mission has always been translated in hardly changing geopolitical and territorial visions. Only the Vietnam War and its aftermath suppressed the idea of mission and the formulation of territorial commitments, at least for a while. It also fundamentally changed the public's readiness to approve of large-scale deployment of troops in foreign countries. American interference, however much coupled with military action, became a way of demonstrating the ability to accomplish a military goal (credibility) rather than the rolling back or containment of an enemy. The same happened with Vietnam in the public mind: it became a fictional world for American supermen. Neither reaction to the Vietnam failure contributed much to knowledge about other societies and political systems. Their attributes remained largely a product of the American symbolic system.

Foreign policy cannot rest entirely on responses to local events in one's own hemisphere. *Sometimes a real war arises that threatens the stability of an entire region.* This happened with the Gulf crisis, which was initially difficult to accept as a *casus belli* for both American public and policy-makers. One week before the Iraqi army invaded Kuwait, US ambassador April Glaspie told Saddam Hussein that the conflict between Iraq and Kuwait was 'none of our business'. That was more or less the message and undoubtedly the mood in the US government. Some days later the editorial staff of *Time* magazine, still unaware of the developing drama, prepared an article on the rising tensions between both countries which appeared under the caption 'Business' (*Time*, 6 August 1990). The likelihood of rising oil prices was apparently the main point of concern. In the next issues, however, *Time* corrected itself by reporting about the conflict under a headline with which they could hardly go wrong: 'The Gulf'. Both facts reveal in their own way how reluctant America was to face up to a possible war and to take sides.

The Bush administration had just decided to concentrate on the national economy, severely ailing after Reagan's careless dealing with the budget. The demise of the communist system in Europe seemed to offer bright prospects for an extended period of quiet attention to business and to new global threats such as environmental pollution and nuclear overkill. 'Good faith [in other nations] can be a spiral that endlessly moves on', declared George Bush hopefully in his inaugural address.[39] A new military threat at a sensitive spot like the Gulf was unwelcome to say the least. It is hardly surprising that a majority of Americans preferred soft reprisals such as an economic boycott to a military adventure.

The outcome of the Gulf War, with very few American casualties and Saddam Hussein maimed, was a great relief to the public. It was the first untainted military success since the Second World War. However, war with Iraq could not satisfy the requirements of a Great Purpose. Policing the world (making the world safe for democracy), although an appealing goal, still turned out to be very costly in view of national economic conditions. Furthermore, Panama's Manuel Noriega, Saddam Hussein and the Somali warlord General Aidid made poor surrogates for Hitler, Stalin and Khrushchev.[40] The euphoria after the Gulf War very soon subsided, although few political commentators would have accepted immediately after the victory that it could not ensure Bush's re-election.

What the Vietnam War or Kissinger's geopolitics could not produce alone, the demise of the communist bloc did: the American foreign policy elite basically accepted the multi-polarity of the world in 1992.[41] They still saw the United States as a superpower, but this was less likely than ever to lead to new 'spheres of influence' thinking. In one respect this may be reinforced by an internal change or cycle that has occurred in the United States, that is, the new generation of decision-makers who have acceded to power. This generation has no personal memory of the Second World War. As Henry Kissinger states in a recent article, 'It has never established the kind of European ties that bound an earlier generation to the Old World and it accepts the Vietnam War as its epiphany.'[42] Bill Clinton is the first president who satisfies this profile and also embodies its inherent paradox. By eluding his military conscription as a young man he became the embodiment of the denial of the postwar world order but also of the American mission. The search for a mission will not subside and the new generation of American opinion-leaders will undoubtedly have to face tantalizing questions about the nature of the new frontiers.

PERIPHERAL DIGNITY AND PAIN

(Argentina)

GOD AS AN ARGENTINIAN

On 24 January 1904 Sir Halford Mackinder delivered his celebrated speech 'The geographical pivot of history' at the British Royal Geographical Society.[1] In that lecture he expounded his vision on the balance of land and sea power, which later became known under the catch phrase 'Heartland theory'. The speech dealt particularly with the significance of the Eurasian continental core. If a powerful state should succeed in dominating both this core – the 'geographical pivot' – and the margins of the Eurasian continent, then, according to Mackinder, the balance of power in the world would be severely upset. In that event we would face, for the first time in history, a 'world empire', an absolutely hegemonic state. Already, Mackinder anticipated a German-Russian amalgamation or treaty. Several forms of counterbalancing coalitions of sea powers were considered in the lecture. The advantages of South America were valued at a premium. Mackinder considered the continent as of decisive significance in a global struggle for power. In the discussion following the lecture one of the speakers even forecasts that South America would experience the same development in the coming half-century as Japan had shown in the preceding fifty years.

Of all the South American countries, Argentina made the deepest impression as it entered the twentieth century. Barbed wire, refrigerated ships and new methods of cultivation lay at the root of an apparently unlimited growth in agricultural production and exports. Argentina became the world's largest producer of maize, linseed and meat and the second largest producer of wheat. For a short period the country seemed capable economically of overtaking the United States. Ex-President Carlos Pellegrini anticipated in 1906 that the race between both countries would be decided in Argentina's favour by the end of the century. There was money, a lot of money: enough to allow the inhabitants of Buenos Aires – the Paris of South America – to enjoy the highest forms of European culture. Prestigious companies like Diaghilev's Ballets Russes with the miraculous dancer Nijinsky performed in Buenos Aires (1911), although some of the members of the ballet sneered at 'le pays du nègre' (making no distinction between Brazil and Argentina).[2] Travelling in the opposite direction, wealthy Argentinian families went to Europe – cheaper than Buenos Aires – to lead a life of leisure.

The half-century between 1880 and 1930 was a period full of hope. The future looked so rosy that the expression 'God is an Argentinian' was coined. The zest for modernization in many territories provoked a broad and fertile orientation towards the outside world, particularly towards Europe. For art and literature France provided the model, for opera Italy, for industry Britain, and for education people looked to an unspoiled newcomer, the United States.[3] The most youthful and futuristic state, and the pre-eminent representative of scientific progress, however, was Germany. Military advisors from Germany were invited to manage the modernization of Argentina's army, while Argentinian army officers went to Germany for instruction. The first director of the Military Academy was a German (Alfred Arent, a pensioned Prussian officer) and Germany became the exclusive provider of weapons.[4] The navy was another story. Here Britain with its maritime superiority provided the model. In this way even the separate identity of the two military services, which was to develop a rather damaging rivalry during the Malvinas (Falklands) War, was cast according to such national differences. German influence would make Argentina one of the few places in the world where geopolitics, in spite of its defilement by the Nazis, remained a practised and valued subject.[5]

Argentina did not in the end become the great ally of the sea powers Britain and the United States, the function it had been assigned in Mackinder's scenario of 1904. In and between the two world wars, the Argentines sympathized with Germany and with Italian fascism. Nor did Argentina become the economically and militarily strong state that the gentlemen at the Royal Geographical Society had envisioned. The comparisons with Japan and the United States likewise turned out to be inappropriate, for both Argentina and South America as a whole. Some time in the twentieth century something went wrong with the promising countries of the southern cone of South America, and with Argentina in particular.

Christopher Leland has called the generation of Argentinian writers born at the turn of the century, who made their names from the 1920s onwards, 'the last happy men'.[6] These writers grew up in a society ruled by hopeful and positivist visions, but during their lifetime social developments took an unfavourable turn that would never evoke the same prospects again. What went wrong?

Apparently – and according to the standard interpretation – it was the worldwide economic crisis of 1929 and after that caused the end of Argentina's prosperity. But that historical fact cannot sufficiently explain why economic depression in the 1930s was followed by stagnation that continues to this day. Other countries that lived through the same economic misery of the Depression did not show a similar stagnation.[7] And if there are more specifically Argentinian political and cultural factors at play, they should also, apart from explaining the process of stagnation and decline, be able to account for the period of successful liberal capitalism at the turn of the century. These are the elements of what has been called the 'Argentinian enigma', or, in Naipaul's words, 'one of the mysteries of our time'.[8]

If there is one feature of social change in Argentina that definitely was caused by the Great Depression, it is the break with 'liberal-bourgeois' political tradition. This was a tradition that was liberal only in its economic views; on political questions it was rather conservative. It was a system with little repression, which nevertheless for a long time depended on a small elite and, when necessary, electoral fraud. Through lack of interest, or put off by bureaucratic procedures, many first-generation immigrants remained without Argentinian citizenship and consequently without voting rights. In spite of these restrictions, most immigrants at the bottom of the social ladder could nevertheless broadly accept the system because it offered opportunities for social mobility and improved general prosperity. In any case, most Argentinians experienced the political climate as 'liberal' at the beginning of the century.

> Here we have neither castes, privileges, closed classes, a feudal complexion, nor historical injustice . . . that would prevent individuals' free fulfilment . . . according to their aptitude, their activity, their intelligence, and their work . . . In this great democracy in the making are no paupers . . . The impermanence of individual bad economic circumstances is . . . almost the rule among us. The rich of today were poor yesterday,

went a commentary from 1912.[9] People did not worry too much about the imperfections of the present because everything was changing and society seemed to be developing inevitably towards a more liberal-democratic order in the future.

Around 1912 the political system indeed appeared to evolve into a more representative type, owing to the institutionalization of an opposition party, the Unión Cívica Radical, or Radical Party, which acceded to power in 1916. 'Representative' is putting it strongly: the electorate in the most important electoral districts still only consisted of 10 per cent of the population, or 30 per cent of the working male population above 18 years. On 6 September 1930 the rule of the Radical Party under President Yrigoyén was terminated by a military coup that employed little violence. From that day military governments and dictators ran the show in Argentina, until the Radicals resumed power in 1983.

That the shock of the Great Depression should lead to a temporary movement in politics away from democracy is not very surprising, considering the world situation at the time. The question, however, is why did Argentina not return to a stable liberal political order when the economy stabilized? In fact, the country initially recovered much better from economic crisis than did comparable countries such as Australia. Urban unemployment in the 1930s never rose above 5 per cent. Political instability is an almost unavoidable consequence of economic adversity, but no other sound causes can be indicated for the ultimate reversal of Argentina's economic development than the change in the political system itself.[10] To explain the tenacity of the new political culture after 1930 it is necessary to search for deeper causes in Argentina's history. It is surprising that these factors did not interfere with the liberal system of the preceding half-century. We know what happened – an economic crisis and a coup – but the important thing to illuminate is why this preceding period did not give enough confidence and security to fall back on?

THE ARGENTINIAN IDEA OF FREEDOM

Practically since the first steps towards independence in 1810, Argentinian society was characterized by a strained relationship between state and citizen. Argentina's history is in any case characterized by disunity and tensions, a feature which constitutes one of the main differences with a comparable country such as Australia. The dramatic moments and confrontations that Australian myth vainly seeks in its own national history, abound in that of Argentina: the struggle between liberalism and the colonial (Spanish) legacy of *caudillismo* (personal dictatorship with its personal favours and obligations); the war with the Indians, ending in the *campaña del desierto* (desert campaign) in 1878; the conflict between the provinces and Buenos Aires about the unitary state; the struggle with British (financial) imperialism after the Great Depression; and finally the undeclared war between state and citizens, with the terror of the 1970s as its low point. These facts are rarely missing in contemporary Argentinian political discussions or in any essay on national identity.

The golden half-century between 1880 and 1930, the triumph of liberalism, took warning from the dictatorship of Juan Manuel de Rosas (1829–1852). Rosas, a successful businessman, became the personification of the reactionary Spanish heritage, the state of barbarism, which, according to early proclaimers of liberalism like Domingo Sarmiento, the country had to cast off by taking Europe as an example.[11] Sarmiento was not thinking of the Mediterranean countries, of course, but of countries where the values of the Enlightenment and the bourgeois revolution had been established, that is, northwestern Europe. However, liberalism, as the antithesis of *caudillismo*, produced the kind of social behaviour that, according to many critics, had less affinity with the aim of general freedom and justice, and more with freedom of the oligarchy. Ambivalence about implementation of the ideal of freedom in Argentina can indeed be heard in the words of one of the great nineteenth-century proponents of liberalism, the political writer and diplomat Juan Bautista Alberdi:

> Liberalism as a form of respect for the opinion of dissidents is difficult to grasp for Argentine liberals. The dissident is enemy, the existence of different opinions means war, an enmity which justifies repression and death.
>
> The Argentine liberals have a platonic affair with a deity that they neither have seen nor know. Being free for them does not mean that one controls oneself but others. Taking control of the state is what their entire freedom implies. Dominating government constitutes their entire liberalism . . . The freedom of the other, they say, is despotism: the government is our power, our true freedom.[12]

Sarmiento was one of those liberals against whom Alberdi levelled his criticism. But even this convinced advocate of liberalism could not, in due course, suppress his own ambivalence about the freebooting kind of liberalism that had developed in Argentina. The waves of immigration that should have raised Argentina to a European level of civilization and development, appeared incapable of invoking a collective national feeling.

Classic Argentine literature again and again shows the breach between state and citizen.[13] Almost all protagonists, even in the novels and stories of the 'happy generation', suffer abuses of justice, persecution or humiliation at the hands of civil servants.[14] Heroes are always enemies of state justice. The 'lawful order' represents a universe that the hero tries to escape. The myth of the *gaucho*, reanimated by José Hernández' epic poem *El Gaucho Martín Fierro* (1872), comfortably fits in with this aspect of Argentine liberalism. In spite of its apparent individualism, it represented an attempt to emphasize an Argentine identity which actually rejects the liberal propaganda for the European model of progress.

In the poem, Martín Fierro, a *gaucho*, is completely immersed in the happiness of his everyday life, family and work. During a feast the *gauchos* present are compelled to join the army in the war against the Indians. The military equipment they are provided with is (by *gaucho* standards) inferior and they do not get any payment. If Fierro complains about these conditions he is due for a brutal corporal punishment. He chooses to desert and returns home. There his wife, in order to stay alive, has joined another man. His children have been set to work. Deprived of home and family Fierro starts to wander and from sheer necessity he falls into crime and even emotional coldness. He humiliates a black woman and commits murder. When the militia finally gets hold of Fierro, he is released by one of its members, an ex-*gaucho*. The two of them flee across the border, to the Indians, hoping for a future in which a '*criollo*' will lead the country instead of a Europeanised liberal.

In the *gaucho* could be recognized a remnant of the Spanish conquistadors, a culture which had survived at the fringe, on the seamy side, of civilized society. Actually *gauchos* were ranch hands, sometimes nomadic, sometimes settled, with legendary skills in the field of breaking wild horses, tracking and fighting. In myth they were the counterpart of Karl May's *Westmänner* in North America: courageous, tolerant, defenders of the weak, keeping their word but always cheated by those representing progress and materialistic ambition. The myth constituted a clear counterpoint to the liberal revolution that had already swept away the real *gaucho* around 1880 but appeared unable to fill the vacuum in Argentinian identity.

If one considers the nature of current complaints about the Argentinian mentality, something must have changed fundamentally. Since the re-establishment of democracy, Argentinian critics have denounced a trait in the national character that turns to the state for the solution of all problems. This attitude does not, however, imply solidarity with the state or with the national community, one respect in which one might establish a continuity with the past. In everyday life citizens associate the state not with authority but with power, not with justice but with entitlements. Those who are not directly submerged in illegality – like the urban guerillas in the 1970s – use politics and personal channels for place-hunting. Politics becomes a means of getting ahead without personal merit, of carrying off a pension or another acknowledgement. The diplomat Archibaldo Lanús opines that this 'politicization of society' (meaning, improper use of politics) has curtailed the role of intelligence in everyday life and blunted the incentive to achieve.[15]

Since 1880 domestic politics had dealt with questions of procedures and power, rather than with policy content. The idea of a loyal opposition was never taken seriously. The party in power, before 1916 the conservative oligarchy, attempted to silence every dissenting political note. The 'unicato', all power concentrated in the hands of one person, the president, was the norm. Under such circumstances being in opposition, as the Radicals were between 1890 and 1912, could hardly develop into anything other than a conspiracy. Nonetheless, the oligarchy admitted the Radicals to government after their electoral victory in 1916. It seemed that a certain institutionalization of democracy had started.

It is never obvious that an elite is willing to share power with another group. There were several reasons why the Radicals were allowed to penetrate to the centre of power. First, their party programme contained no revolutionary goal. In contradiction to what the name suggests, the Radical Party was a real middle-class party with the same economic aims as the oligarchy of wholesalers and landowners. It differed only in its emphasis on institutional change, such as the extension of voting rights. Second, the widened representativeness of state power – however illusory it might have been – reinforced the political front against anarchists and revolutionary socialists, who had emerged at the turn of the century. As a movement anarchism did not pose a real political (electoral) threat but its violent expression scared the elite. Finally, easily gained prosperity had induced a certain political inertia. It was thought that each wave of dissatisfaction could be bought off with the revenues of new growth. That economic growth would continue was never doubted. It was believed that providence had predestined Argentina to produce without effort, or, as recounted in Alberto Gerchunoff's novel *El Hombre importante* (in an almost perfect, albeit ironical, echo of Ayarragaray's serious words of 1912):

> The agricultural production is simple and restricted to the spontaneous wealth which God has put in our land. We export wheat, corn, linseed, meat and linen. Our collective life is not confused by the opposition between hostile and exclusive interests, by various industries, moral questions about family life, established coalitions or an overstrained and hypersensitive consciousness. What is wrought in other countries by skilled and experienced leaders, Holy Providence takes care of in our country.[16]

Seen from this perspective, there was indeed little reason to worry about the absence of a national policy discussion or the necessity to start a dialogue between political insiders and outsiders. But precisely this situation, in which the rules of the game were not yet crystallized out, while an apparently irresistible process of institutionalization of new interests had started, created the conditions for the conservative panic of 1929–1930. The profound vision of the state as just an extension of particular interests also contributed to this situation. The coup had to safeguard the interests of the great wholesalers and the agrarian elite. Although it also brought a military government, this was not in itself the cause of the protracted economic stagnation.

The question of whether Argentinian political culture, through its own dynamics, provides a key to the changing fate of Argentina has been answered in the affirmative by Carlos Waisman. His key is the special course of the democratization process between 1880 and 1930. As a result of the reassuring prosperity and immigration, the democratic institutionalization was too sluggish to create the political consensus or willingness to compromise that was necessary to give a pragmatic response to the economic crisis in 1929. On the other hand, the democratic surge was strong enough to frighten the conservative elite. There are, however, other interpretations. Some authors are inclined to put more emphasis on the nature and consequences of the Depression of 1929 itself, on self-defeating forces in the national mode of production, or on the grip of foreign (British) capital. The Argentinian economy was never safe, Gary Wynia says. Twice, in 1890 and 1919, the oligarchy succeeded in riding out a recession, expecting that the losses would be compensated after the return of growth. However, in the 1930s, these options were forgone.[17] The fact that the performance of Argentina's economy in the 1930s was comparatively good is ignored in this argument.

METAMORPHOSIS OF THE ARGENTINIAN STATE

According to Waisman, the basis for the real reversal in Argentina's development was laid much later: after the coup of 1943, and particularly during the rule of Juan Perón (1946–1955). Only by then was the internal orientation of Argentina's economy definitely settled, inasmuch as the connection with the world market was broken and industrial competitive power started to diminish. The Great Depression did have a delayed effect as a source of images of the development of the world system, but only in combination with the later experiences of the Second World War.

It cannot be denied that the Depression had immediate effects too: first, most countries in the world were raising tariffs to protect their national industry – as did Argentina – and, second, the liberal model of economic development became severely criticized. Or, more explicitly: an increasing number of people and opinion-leaders believed that there was no future for liberal capitalism (a model which seemed to have reached its height in the nineteenth century), only for really new models of society like communism or fascism. Suddenly world events seemed to confirm the view that capitalism was nearing its end as an historical stage. This is what occurred in Argentina but in Europe, too, people could not get away from this impression, the liberal democracies included.

If indeed lessons are to be learned from foreign countries, then events in countries that are culturally closest will make the deepest impression. For

Argentina these countries were Spain, culturally similar and the source of many immigrants, and Italy where most new immigrants came from. The Spanish Civil War was followed with much attention and concern in Argentina. The message seemed to be that liberalism and communism could only produce chaos. The fascist revolution in Italy, on the contrary, proved that power and stability could be attained another way.

Later, when he acceded to power, Perón would use his personal experiences in Italy to bolster the Argentinians in rejecting liberal democracy. Perón had been sent to Italy by the army in order to acquire military knowledge and to study the regime. He also followed a course in fascist political economy at the university of Turin.[18] He never made a secret of the overwhelming impression which Benito Mussolini made on him during his stay. For long afterwards (in exile during the 1950s and 1960s) Perón sustained his claim that the fascist model was superior to capitalism and communism. His admiration was shared by many Argentinians, although few of them were inclined to conclude that the Italian system had to be copied in Argentina. For this it was judged to be too bureaucratic and not sufficiently spiritual (or religious), the same was certainly true with respect to Germany, which was also admired. 'The new Germany is viewed as hostile to culture . . . because of its supposed threat to the Catholic church,' the Nazi diplomat Richard Meynen wrote from Buenos Aires to his superiors.[19]

In tune with the reversal in the public mood after 1930, national history began to be encoded in a different way. The social unrest that had followed previous recessions, particularly the events immediately after the First World War, now became evidence of an increasing subversion of the established order by revolutionary, foreign-inspired, forces. The immigrant suddenly became a potential terrorist, member of a 'phalanx of fanatics who came from different parts of the world with the deliberate purpose of disputing what is ours.'[20] Elements of xenophobia had also surfaced during the turmoil of the *Semana Trágica*, the tragic week of January 1919. Initially this involved an expression of labour unrest, following disturbances of the world market after the war. Argentinian exports had diminished severely. Wages were subjected to great pressure (a decrease of 22 per cent), culminating in strikes in 1919. The state reaction to demonstrations was violent and motivated by reports of anarchist attacks and the revolutionary events in Russia. Hundreds were killed, not only through violence of the official police. Nationalist vigilante groups ('sporting clubs') were involved, attacking everything that could be associated with the Left (communism) or with foreign countries. The life and property of residents in the Russian-Jewish quarters of Buenos Aires got suspiciously little protection against such attacks and later claims were hardly taken seriously.

These emotions and fears, which subsided after the return of economic growth, were rekindled in the 1930s and 1940s. Perón used the fear of the 'dangerous class', migrants without social ties and workers, to sell his corporate state model to the elite. The message was that the revolutionary threat and labour unrest could be bought off with a model in which the working class sacrificed part of its freedom and the elite part of its wealth.

This Peronist principle has never been wholeheartedly accepted by the representatives of agrarian and industrial big business. It was more that in Perón they recognized one of those gang leaders who extort money from businessmen in exchange for 'protection'. The dynamics of the state as a force detached from particular interests had progressed too far to restore the pre-1930 liberal order.

Protection of home industries and a cult of economic independence, were the basic components of this model. 'No market can replace the internal market. Regardless of how big, how beautiful they may be, because [other markets] are always aleatory, never safe ... If something is left over, we will sell it somewhere, but the Argentine people come first'.[21] That was the message of Perón when he became president. In these words reverberated not only the shock of the Depression but also, probably more importantly, the experience of the Second World War.

The German advance against militarily strong states including Britain, France and the Soviet Union greatly confused the political elite. Together with the established respect for German military skills, this resulted in the expectation that a Europe ruled by a totalitarian Germany was a serious possibility. Perón had come to know Germany during his stay in Europe as a 'gigantic machine which functioned with marvellous precision and in which no small part was missing'.[22] A picture of a postwar world of 'autarkic continents' started to emerge. A liberated but destroyed Europe, would not be able to buy Argentine products for a long time to come.

These thoughts inspired a strategy that aimed at economic independence, perhaps not initially as an economic credo but certainly as a means of guaranteeing national security in time of war. A policy of careful neutrality was maintained, which consequently reinforced Argentina's isolation. Its neutrality was understood as a hostile act in the United States, particularly when, in accordance with custom rather than as a token of any closer relationship, Argentina started to buy weapons in Nazi Germany. A wartime economic and military boycott by the United States and Britain and, after the war, a boycott by their European allies, was the result.

Given these conditions, a country has little option but to resort to its own economic resources and market. Yet, there are reasons why Argentinian politics should not be considered as an unresisting plaything of the international system. Waisman calls the weakness of Argentina's will to take a more critical view toward Germany unseemly. Relations with Nazi Germany were indeed broken off in 1944, under American pressure and after it had become clear who would triumph, but an official declaration of war on Germany only came a month and a half before the fall of Berlin. A demonstration in Buenos Aires to celebrate the event was prohibited.

A further reason to ascribe developments to deliberate Argentinian choices and cherished opinions is that the inward orientation of the economy continued after the economies of European countries began to revive at the end of the 1940s. Argentina had already voted for Perón. By the end of Perón's rule, in 1955, the legitimacy of state authority was undermined to such an extent that a return to free-market principles had become impossible.

The Peronist system had created a severe trade deficit. Agrarian exports had decreased as a result of increased domestic consumption, while the protected national industry was not competitive enough to restore the balance of imports and export. When wage levels could no longer be guaranteed, the legitimacy of Perón's model evaporated. The diverse social interests got into a stalemate characterized by frequent changes of power and undemocratic regimes.

NATIONAL SELF-ANALYSIS

It is characteristic of the despair felt about the decaying state and economy that Argentinians were unable to denounce the political course of the country after the war, that they fundamentally did not want to do so, and even recalled Perón in 1972 as the saviour of the nation. From exile in Spain, Perón promised all the interest groups that appealed to him exactly what they wanted. He promised law and order to the military, a sound climate for investment to business, a social revolution to the rebellious youth, and national unity to all. Everybody knew it was all rhetoric, says Gary Wynia, but they went on hoping, against their better judgement, that the promises would be fulfilled.[23]

The dictator returned, accompanied by the embalmed body of his late wife Eva Duarte, who was worshipped by the people as a saint. 'That . . . is a story I could *never* write,' said an ironical Jorge Luis Borges, the Argentinian writer who, nonetheless, became famous for his fantastic stories.[24] Borges belonged to the sceptical elite – the last happy men – who had always looked upon Perón with contempt. He had already established in 1930 that the Argentinian, unlike the North American and European, does not identify himself with the state, which 'can be attributed to the fact that governments used to be bad in this country or to the general fact that the state is an uncomprehensible abstraction.'[25] 'Lack of imagination': so Borges characterizes the Argentine mind, but this ailment did not prevent the people from identifying with a myth – the one of Juan and Evita Perón – and it did not prevent the myth from enduring even after Perón's final failure.[26]

In economic analyses of Argentina's predicament, the valuation of Perón's rule is the reverse of the judgement from those who focus on the problem of national identity or the fate of the poor. The economist's judgement is usually crushing. Shortly after the Malvinas/Falklands War, several Argentinian intellectuals felt called upon to publish their reflections on the Argentinian national character and how it could be restructured to make a more satisfactory society. For one of them, the novelist Diana Ferraro, Perón was the first leader who succeeded in bridging the many oppositions: between the different economic classes, between the coastal region (Buenos Aires) and the pampas and, not least, between men and women.[27] He was the first to assign a political role to women and under his rule women's suffrage and other forms of emancipation were implemented. After his fall the general situation with regard to democratic rights in Argentina deteriorated, concludes Ferraro. As liberal economists consider Perón's rule as the decisive event in Argentina's decay, nationalist minds, like that of Ferraro, consider Perón's decade (1946–1955) as the culmination of a development which has been piteously frustrated by other – 'imperialist' or 'materialist' – forces.

For Diana Ferraro the great crisis of 1929 actually continues (in 1985), even in economically successful countries. She considers capitalism and communism as equally poor and equally materialistic answers to the question of the organization of the state. Argentina would do better to search for other models. She recommends turning to the national spiritual roots that have been overgrown by five centuries of materialism, or to the 'forces of nature' that were present in the Indo-American civilizations. With such opinions the authoress sides with the many other voices which have, in the course of Argentina's history, translated feelings of frustration about confrontations with the outside world into an emphasis on *criollo* identity. But now the idea is that this culture involved certain principles that have become obliterated elsewhere – in Europe – and that might give Argentina a new mission in the world.

The idea of a 'third road' for Argentina that is neither capitalist nor communist is nothing new. The idea reverts directly to Perón. But while Ferraro considers Perón's views on an economically and militarily unaligned position as unrealistic in the situation after the Malvinas/Falklands War, she suggests a new kind of Peronism on the basis of post-modern and post-materialistic values. In the field of international relations this would mean a farewell to Perón's anti-Americanism and the 'ultra-nationalistic play' of the generals in the Malvinas/Falklands. A country that has solved its own identity problems should not be afraid of international obligations and loyalties. What, however, could this identity or mission, that should be able to bring to an end the series of piteous failures in the field of international politics, be? Ferraro's treatise vaguely hints at an Argentina that, as in the earliest times of Spanish colonization – the viceroyalty of the Río de la Plata – will resume its role as 'guardian of the South'. In the current international order this would indeed require a rapprochement with the much criticized United States, rather than national isolation.

As a consequence both of German influence early in the century and of Argentina's geographical situation, geopolitical reasoning has never been an exotic element in Argentine political discussion. Of course, it appeared in a more detailed way in the discourse of generals and foreign policy specialists than in public discussion. But the publication of geopolitical treatises and the presence of people in high government positions who communicate such ideas is in itself a significant feature of South American countries like Argentina, Chile and Brazil. The absence of the burden of war experience in the Second World War and the experience of an unfinished environment, both internal (Patagonia as colonization area) and external (Antarctica, the oceans) explain this persistence of geopolitical thinking. Indeed, although the outcome of the Malvinas/Falklands War may have boosted the public's distaste for geopolitical thinking, notions of being unfulfilled, in danger, and abused, have survived in the feelings towards the international environment. This actually continues the Argentine self-image which after the 1940s shifted from one of a 'new' European-style country to that of a typical 'Third World' or Latin American country. The feeling of national decay – El malestar Argentino in Diana Ferraro's words – penetrates all domains of life. According to Jack Child, who published several studies on South American geopolitics, this was also the message of geopolitical writers, most of whom depicted Argentina as a country subjected to frequent aggression from its neighbours (Chile, Brazil) as well as from foreign powers (Britain, the United States). In the words of one general:

> It is painful to say it, but Argentina is perhaps the only country in the world which, throughout its history, from the moment of independence to our days, has given up territory, as a consequence of the fact that our ruling class has not considered space valuable as a power factor; it has not borne in mind that to diminish the space of a Nation is to diminish its power.[28]

Such feelings undoubtedly underlay the unanimous public approval of the Argentinian occupation of the Malvinas/Falkland Islands. At the same time the frustrations about the unexpected outcome of that action, first the outbreak of war with Britain and subsequently the military defeat, shifted the pursuit of national pride to 'national projects' in economic and cultural fields rather than international deeds.[29] The traditional opposition between the nationalist spirit, drawing on a Christian-Spanish heritage, and the liberal, cosmopolitan spirit is still manifest in such issues. While Diana Ferraro searches for immaterial inspiration in a distant time before the Reformation and Enlightenment, the reflections of a liberal like Archibaldo Lanús result in praise of the cosmopolitan 'model city' (ciudad guía) Buenos Aires as the nation's trump card.

NATIONAL MYTHS AND INTERNATIONAL PERCEPTIONS

Two factors seem to underlie Argentina's interpretation of its place in the world, a world image which has resulted in political decisions with far-reaching consequences. The first is the lack of national cohesion resulting from Argentina's typical history as a country of high immigration, and the second is its peripheral position with regard to Europe, the world with which, desperately searching for its own identity, it nevertheless remained closely connected in both a cultural and an economic sense.

This lack of national cohesion was provoked by the individualist mentality of most immigrants who crossed the Atlantic to 'make it to the top in America', and by the cultural background of immigrants' home countries, where a 'civic culture' involving individual feelings of responsibility for society was weakly developed. The fact of this traditional contrast between the political cultures of the countries of southern and northwestern Europe is generally accepted. Labour unrest in the first decades of the twentieth century in Argentina, lack of tolerance for other groups and opinions, and the frail democratic disposition of national leaders, all have to do with these origins.

Negative judgements on democracy as a system have been recorded frequently from representatives of the Roman Catholic Church or from others invoking religious principles. The priest Meinvielle (1903–1973, a prolific writer and champion of the rightist cause) called fascism 'less totalitarian' than liberal democracy because democracy would suppress the 'spiritual force' of a society and deliver society into the hands of a squandering bourgeoisie.[30] Even more recently, and certainly since the democratic change of 1983, voices have been heard identifying democracy with 'that horrible legacy of the French Revolution' or with a 'Zionist conspiracy'.[31] Such opinions are not necessarily representative of the majority of Argentinian citizens but their tenacious occurrence and shameless expression in public suggest a range of options for serious political debate which differs essentially from what mainstream political culture in postwar Europe allows.

The combination of Argentina's geographical position, the disruptive effect of economic cycles since the end of the nineteenth century, and the experience of two world wars from the viewpoint of an outsider, contributed to the choice of betting on an unaligned, and ultimately marginal, position in the world. The gamble failed, but when the Argentinians realized this, the European schism of the system of communist and capitalist blocs again seemed to proclaim the bankruptcy of Western culture and to emphasize the inevitability of an independent, Argentinian road. The Argentinians reassured themselves by magnifying every trace of disunity in the rest of the world – communist against capitalist, Catholic against Protestant, Britain against the United States – thus also helping to downplay domestic tensions and conflicts.[32] There was certainly a more than indifferent feeling towards the outside world, not in the sense, once expressed by an Argentinian, that 'we are not neutral, we are against everyone', but in that of an overpowering negative impulse: the disgust and fear of communism.[33] This attitude linked up with the intense reaction against political agitation and labour unrest that had characterized Argentine society earlier in the century. The new anti-communism, however, could not get the same expression as in Europe. Whereas in the countries of western Europe the enemy could be represented as a distinct power immediately threatening their borders, Argentinian politics had to concretize the danger on two fronts, either as an internal enemy or in a dramatic scenario about a strategic crisis in the South Atlantic.[34] It is part of the tragedy of Argentina's history that the generals who acceded to power in 1966 plunged the country into a fateful phantasmagoria on both fronts. They first dropped the distinction between internal *instability* and external *security*.[35] Thus the military got entangled in a 'dirty war', a phrase with many connotations. It not only meant a war against an enemy who does not fight openly, but also the sabotage of national unity, a civil war, including its inevitable human rights violations, and finally a type of war which can never be won with purely military 'standard operating procedures'.

However, it was the defeat in that other war that 'could never be won' (as the Argentinians later declared emphatically), the Malvinas/Falklands War, that most clearly revealed the derailment of the military machine and broke its power. During the 1970s military strategists in South Africa and Argentina had painted a future in which the South Atlantic could become a communist lake as a result of the opportunities offered to the Soviet navy in Africa. The Soviets were thought to be particularly interested in the mineral treasure-house of South Africa and the potential wealth of Antarctica. Antarctica also evoked the idea of the strategic importance of the Cape Horn route if the Panama Canal should become blocked by a communist insurgency in Central America. These ideas never succeeded in persuading the United States to support actively a South Atlantic Treaty Organization. Old distrust of Argentina's unaligned position and the rivalry between potential partners like Argentina, Chile and Brazil, and other factors were impeding the materi-alization of an international coalition in this area. In the Malvinas débâcle Argentinian nationalist obsessions constituted a principal factor inciting the United States to dissociate itself from Argentina's cause and other, neigh-bouring, countries to hold back from offering real support in spite of many fine declarations of Latin American solidarity.

The South Atlantic offered an opportunity to shape Argentina's identity, but this intensified competition with other South American states that could boast of similar strategic advantages. Anywhere else in the world, a 5000-kilometre border, like that between Argentina and Chile, would have yielded some tensions about disputed territories. However, no border conflict kindled Argentinian emotions so much as the line of the border through the Beagle Channel at the southern tip of the continent. Actually it concerned three tiny, barren islands (Picton, Nueva and Lennox), but the underlying cause of the emotions was fear of its effects over a much wider territory. As General Villegas stated: 'If they take the sovereignty of the southern islands away from us, we will, sooner or later, lose our rights in the South Atlantic and will compromise our claim to the Antarctic sector.'[36] In 1978 Argentina and Chile almost went to war over these islands but arbitration by Pope John Paul II prevented further escalation.

The disappearance of the Soviet threat meant that the argument depicting Argentina as a lonely crusader for the Western Christian cause lost its force. Although claims on the Antarctic sector (and the Malvinas) remain an important national interest, the sad state of the national economy and the acceptance since 1986 of free-market principles under the 'Peronist' Carlos Menem have thrown Argentina onto its own resources and more pragmatic external relations. Yet, national politics still rest on the Peronist legacy of a marriage of 'deep loyalty and pragmatism, charisma and expediency'. Peronism 'still tends to behave in the government as if it were an entire political system unto itself.'[37] The Argentinian mind can easily switch to new national goals that satisfy the need for national dignity but do not particularly serve the national interest in other ways. For the time being it will be difficult to derive this identity from a special geopolitical position or task in the wider world. The most urgent need is to correct the legacy of a politics that assigned to Argentina a peripheral position as a place of honour.

WANDERING IN CIRCLES

(Australia)[1]

In 1992 Australia's Prime Minister Paul Keating caused an uproar in the British press by touching the British Queen irreverently during her official visit to Australia. Emotions were high because of Keating's image as someone who shortly before had demanded the establishment of an Australian republic, and an end to flying the Union Jack and participating in the British system of honours. Keating's gesture (in which he offended against protocol by putting his arm around the Queen), seemed a symbolic Proclamation of the Republic in the way Napoleon became Emperor by snatching the imperial crown from the hands of the Pope. Why such susceptibility concerning symbols when Britain's voice in Australian affairs was practically nil already? In Australia the idea of being indebted to Britain, even in a symbolic way, hits an increasingly sensitive spot. This development is all the more remarkable since British citizenship was still highly appreciated fifty years ago. It ranked as a privilege and reason for pride.

'Geography versus history' runs the motto heading the last chapter of a well-known text about the history of Australia's external relations.[2] History has moulded Australians after the image of Europeans, or rather Britons, but the important decisions on security and economy have brought the Australians face to face, in the first instance, with South-east Asia and the Pacific region. Until recently this missing link between head and limbs in the Australian nervous system provoked painful self-examination. Currently Australia seems to acknowledge, with fewer reservations than before, that its future lies in partnership with the states of the Asia–Pacific region.[3] This reorientation has clearly helped the ailing economy, but does it also imply a mental integration? A popular test in which respondents draw a world map reveals that students from secondary schools in Australia and New Zealand still know the politico-geographical details of Europe much better than those of other parts of the world, including their own 'region'. And, in spite of the flying diplomacy of Paul Keating in the region (a novelty in Australian politics) after his takeover in 1992, South-east Asian leaders remain dismissive of Australia's claim to be an Asian nation. One of the sensitive spots is the question of human rights. Even though Australian leaders have decided to keep a low profile with respect to human rights, they cannot prevent the media or other politicians from criticizing the situation in certain Asian countries in the region.

Nobody would demand that the Australians drop their values for the sake of a smooth integration into the region – apart from the sheer impossibility of giving up something that is part of the national character. For a good match between values and reality one needs an identity. Many signals from daily life are evidence that for most Australians the question of national identity is not settled. The correspondent of the *Independent*, Brian Appleyard, even established that this nervous search for national distinction had produced 'the most insanely political nation in the world'.[4] In no other country would so much commotion arise about a person as it did in Australia when the 'national historian' Manning Clark fell into disgrace. It was because the reliability of his historical analysis was questioned, as if the nation's pride in its history went down with him.[5] Parties and personalities are blown up in the media until they assume dazzling proportions that bear no relation to their real interest. This politicization of culture, however, does not indicate a great variety of contesting ideas concerning the world.

THE PROBLEM OF NATIONAL CHARACTER

If there is one Australian characteristic on which many observers agree, then it is the very propensity to avoid ideological reflection. The Australian's pragmatism is frequently noticed. It evokes at least astonishment and some-times negative comment. The British journalist Linda Christmas, a generally sympathetic observer, noticed how during an exciting election campaign in which the Liberals lost their government majority, the goals of the different candidates were hardly a subject for discussion (L. Christmas, *The Ribbon and the Ragged Square*, p. 61). She was told that Australians actually do not relish discussions and ideas and that they do not like grand visions either. Even learned studies of Australia's political culture conclude that 'interests' play a more prominent part than 'ideas'. The only political truth which would interest people is couched in questions like: 'Does it work?' or 'How many are behind you?'.[6] A study of Australian autobiographies establishes a common element in what writers report about their social environment: a strongly authoritarian society, intolerant toward human differences and fearful of deviant behaviour; a strongly class-conscious society that emphasizes materialist and pragmatic values and in which the arts are distrusted or even treated with hostility.[7]

The peculiar history of Australia as a penal colony is always put forward to explain these supposed national characteristics. The first 'migrants' who set foot ashore in Botany Bay were convicts who, as far as they still possessed a trace of civilization, coarsened very rapidly under the influence of the untamed nature and hardships of the first colonization phase. Apart from the bottle of rum, the exiles had only one option to alleviate their existence in an enduring way: working for material rewards. Convicts 'who shall from their good conduct and a disposition to industry, be deserving of favour' could obtain land property rights.[8] Almost from the outset material gain was the dominant reason for existence of the Australian colony, a start essentially different from that of the colonization of North America, where Europeans went on their own initiative to build a more ideal and pure society than European conditions permitted. This idealism and moralism, though strangely mixed with materialism, has left deep marks in the current United States. The long period during which the inhabitants of the Australian colony saw themselves as second-grade citizens of a large English-speaking empire would have encouraged the adoption of materialism as the only safe 'lingua franca'.

A reliable historical account should beware of the dangers involved in using the national character as an explanation of historical developments. Even if we assume a decisive role for collective attitudes and values in keeping up certain institutions, such psychological features may change. Why should a culture not change – in the long run – if social and economic conditions are fundamentally overturned? On few subjects do ideas rely so much on quotations and on echoing the opinions of others as on national character. Likewise, in few intellectual activities does the presumption of properties so easily entice into actually observing the same characteristics as in the description of national or ethnic groups. Even trusted sources like autobiographies may raise doubts because their writers are presumably a selection of persons who are relatively weighed down by the materialism of society, wherever they may live. Moreover, application of the well-known post-materialist/materialist attitude scale does not reveal a deviant position for Australia among the widely diverging scores that can be established even within Europe.[9] The Australian score (for 1988) did not differ essentially from the one for France, the home of *esprit*. Likewise a comparative study of political attitudes in six countries – Australia, Germany, Britain, the United States, Austria and Italy – pointed to Italy and the United States as deviant countries rather than Australia.[10] In many ways, however, Australia reveals an affinity with the Anglo-American type of political culture.

Prevailing public opinion research does not plot the 'meaning-dimension' very well in cross-national studies. Yet, for the time being the most cautious conclusion is that, as regards the preferences and ideals that control everyday life, no essential gap exists between Australians and citizens of other (Western) countries. The comparatively strong pragmatism in politics can be traced back to the absence of 'great narratives' that connect territory, history, national epics and sentiments to each other and to the world at large.

NATIONAL MYTHS

The specific historical development of Australia has generated citizens who for a long period identified with European (British) interests and visions. The primary identification was with Britain, after that one was a citizen of Victoria or South Australia. The third possibility, identification with Australia as a whole, only appeared when Australia started to play a more or less autonomous role in the world. Initially, for the individual Australia was an identity problem rather than a personal asset. The year 1901 can be designated as the start of Australia's independent existence. In that year Australian home rule was instituted and the groundwork was laid for an Australian constitution. The change was not extorted from Britain by violent resistance and irritated demands but resulted from the 'progressive' British insight that decentralization of the empire would preclude many practical problems and relieve its administrative burden. The maintenance of foreign relations remained to a large extent in British hands.

When the First World War broke out, Australia's participation was in no doubt. There was an Australian representative in the war council but no option for or against participation in the war.

> The bugles of England were blowing o'er the sea
> As they had called a thousand years, calling now to me.

These moving words were written by a young Australian soldier who was soon to die at Gallipoli. Australia's toll in the First World War amounted to 60,000 dead soldiers, 20 per cent of those conscripted. The lost battle at Gallipoli on the Bosporus in 1915–1916 was the most memorable event for Australia's soldiers and a rich source of ambiguous national feelings. On the one hand, it seemed to prove the disastrous consequences of leaving decisions to non-Australian commanders. Australians also dissociated themselves from the kind of military conceptions that had turned the war into such a meaningless slaughter. On the other hand, these deaths were of the first victims in Australian history who died for a higher cause. As everywhere in the world, they could provide the construction material for the temple of national identity: heroic deeds for the fatherland.

Only in the Second World War, when Australia's national security was at stake, would the impetus be produced for a real mental embrace of Australia as an independent nation. The awareness of geopolitical vulnerability during the war was the deciding factor in sending the first diplomatic representation abroad: to the United States. After all, events in Europe had made it clear that no longer could the security of Australia be guaranteed by Britain. The gap between Australia and Britain became wider still when the Australian government, against the wishes of the British, withdrew its troops from the Middle East to make it possible to better protect Australia's security against an increasing Japanese threat. Half a century later Paul Keating, in nationalistic mood, referred to this episode in emphasizing how much Australia had been left alone by the British in its darkest hour.

Until now the negative drive, the wish to distinguish oneself from Europe, has been an important element in Australian attempts at self-definition. It proves, in a paradoxical way, the immaturity of Australian identity-formation. The history of harsh struggle in the Australian 'bush' was to provide the same kind of romantic ingredient the Americans found in the 'frontier' and the Argentinians in the *gaucho*. The myth places a sturdy man of the wilds against the degenerated European man of culture, a male against a female ideal. The first image denotes a human being who exerts power over his environment, the second is merely a dependent product of environment and history. But positive as this comparison is meant to be for Australia itself, it arises from a national myth that still remains to be shaped. To have power over one's environment immediately evokes the identity question: to make what?

In the light of such conditions it is hardly a coincidence that one of the most beautiful descriptions of national disorientation comes from a woman (though from the pen of a man) in the novel *Homesickness* by Murray Bail (1964):

> We come from a country . . . of nothing really, or at least nothing substantial yet. We can appear quite heartless at times. I don't know why. We sometimes don't know any better . . . Even before we travel we're wandering in circles. There isn't much we understand. I should say, there isn't much we believe in. We have rather empty feelings. I think we even find love difficult. And when we travel we demand even the confusions to be simple. It is all confusing, isn't it . . . ?[11]

'Wandering in circles' is something one lapses into when there is no clear starting-point or when one takes one's bearings from something that is itself moving. This lack of focus is just as characteristic for the explorers' expeditions into the interior of Australia – without the dramatic interactions or suspense that characterized the reports of voyages of discovery in other continents[12] – as for Australian foreign policy in general. In foreign politics first Britain, then the United States, and finally a kind of national interest, 'standing up for Australia', served as beacons in the turbulence of the world. However, even this last reference point offers a less firm hold than one is inclined to believe and Australian trade policy teaches.

GEOPOLITICAL VISIONS

Hardly a week after Saddam Hussein's troops had invaded Kuwait, the Australian Prime Minister Bob Hawke responded to a telephone call from President George Bush with the promise to send three warships to the Gulf. He acted apparently without consulting the other Cabinet members. A 'rueful triumph for golf course diplomacy', commented one newspaper, referring to the sporty relations between the two statesmen.[13] The event immediately evoked comparisons with the Australian commitment to the Vietnam War, when the Australians had likewise sided with the Americans because they were convinced that fighting communism, particularly the danger from China, was in their national interest. But this conviction was not supported by an explicit Australian political strategy towards South-east Asia. Fear of communism's territorial threat was to a large extent the official motive for contributing to the forces in Vietnam. Maintaining the alliance with America was the underlying aim. The unanticipated decision of the Americans to withdraw from the war embarrassed Australian politicians, who suddenly found themselves in an ideological vacuum without clear national goals to fall back on. The same happened when President Nixon started his political rapprochement with China. In July 1971, the American negotiator Henry Kissinger and the Australian opposition leader Gough Whitlam were both, unknown to each other, in Peking to talk informally about a rapprochement which the Australian government had always passed over both out of conviction and out of respect for the Americans.[14] On his return Whitlam was scorned as a traitor, which, ironically, was not what was to happen to Kissinger three weeks later when the White House revealed his secret mission to Peking.

All these thoughts and frustrations were reactivated by the decision of Prime Minister Hawke to support the Americans in Operation Desert Shield. Once again, it seemed as if the country had jumped into an adventure which could have no proper Australian continuation, an action which would bring unpredictable developments initiated by the United States. The most essential conclusion was that there had been no shortage of moral statements by opinion-leaders – for or against military interference – up until military operations started in January 1991, but there was a lack of *discussion* of both positions in the United States and Britain. Nor had the Australian government undertaken steps to take decisions in concert with the other countries in the region, so there was no regional rationale in Australia's decisions. All the attempts in preceding years to adopt an image of a South-east Asian country seemed to be undone by Australia's unconditional espousal of the American point of view. The Gulf crisis demonstrates, according to James Richardson, that in Australian politics 'despite tendencies toward cosmopolitanism and greater sophistication, there can be relapses into narrow provincialism, closing off a sense of larger issues and of ethical dilemmas.'[15]

Actually the belated acknowledgement that Australia was, geopolitically and economically, an Asian country, was caused by the prominence of a long tradition of global visions in Australian political leaders. Even after the Second World War Australian leaders seemed to believe that 'the proper place to discuss matters with Asians was in London.'[16] Such visions were maintained in spite of the fact that the Second World War and its aftermath had both shown the strength inherent in Asian countries such as Japan and that the European colonial powers were withdrawing from South-east Asia. This process, however, was more than compensated by the expansion of Western military and political influence within the framework of the Cold War. South-east Asia acquired the meaning of a territory that, in the context of global containment politics and Australia's forward defence strategy, had to be defended against communism. The partnership with the United States and New Zealand (cemented in the ANZUS treaty of 1951) and the special relationship with Britain offered a more extensive security guarantee than previous unilateral dependency on Britain. Besides, this arrangement kept Australia within the domain of Anglo-Saxon, white culture. It helped to maintain the cherished aim of white Australia.[17] (The formation of the regional military association SEATO in 1954 did not really change the Australian state of mind. Interaction with other South-east Asian states was a matter of principle rather than practice. SEATO's context of American power was the main significance for Australia.)

The ANZAC pact between Australia and New Zealand (1944) and the ANZUS treaty between both countries and the United States (1951) affirmed the separation of Australia and New Zealand from their wider region and reinforced the idea of Australia as a special and isolated example of Western culture. The feeling of isolation was mirrored in public declarations that the problems of the rest of the world (such as population growth, East–West relations, and peace questions) were simply irrelevant to the Australian reality. This partly explains why a 'narrow provincialism' (to quote Richardson) occurs simultaneously with an inclination to cosmopolitanism.[18]

The memoirs of the penultimate Prime Minister, Bob Hawke (1983–92), provide a special illustration of the mix of these paradoxical Australian attitudes.[19] At the one hand, the overall message of Hawke's reflections is clearly to portray the author as an international leader who has pulled his weight in bringing many international crises to a solution, even the conflict between Israel and the Palestine Liberation Organization.[20] At the other hand, the author indulges in descriptions of small personal contests which he time and again wins (The 'Mr Hawke, we used to hate you, but you were right' style). The peculiar feature of these discussions is that they are primarily described as a struggle between players (like in sports) whereas the intellectual content of the discussions is always secondary. Hawke refers once or twice to his interest in religion during youth, but there is nothing in these memoirs which proves another interest than material problems and power.

One might expect that the shattering of Australia's geopolitical vision, which the Second World War could not bring about, would at least start with the American retreat from Asia and China's new openness to the West. Indeed prime ministers Whitlam (1972–1975) and Fraser (1975–1983) proved themselves aware of the need to establish more equal relations with Asian neighbours. They also started to change immigration policy with respect to an envisaged 'multi-racial society'. However, global political concerns did not immediately retreat into the background. At the end of the 1970s the Soviet invasion of Afghanistan again produced a heightened concern about global security problems which delayed further developments towards regional integration for some time.

The criticism levelled at Australian political culture by its intellectuals very often concerns the lack of an autonomous vision of the world. Australia is depicted as a passive follower of rules which are made and broken elsewhere, leaving the country in a confused state. One might ask whether this criticism accounts sufficiently for the limited range of options which any not very powerful country has at its disposal. A sneering commentator might even suggest that the national syndrome of a 'small' country is not so much its lack of power as the frustration of its intellectuals.

Richard Leaver is of the view that Australian reactions to American policies have not been as docile as many critics claim.[21] At the height of the United States' struggle with the communist 'danger' in South-east Asia, Australia would always have resisted an excessive expansion of American power in the area. When the Western presence in the region began to crumble and the future loyalty of the United States to ANZUS started to come into doubt, the Australian government made great efforts to elicit an American commitment to Australia by offering the United States military bases. Later on, when the

Cold War began to loosen its grip, Australia read the United States a lecture on its lack of purpose in the world. These are all indications of a distinct autonomous policy, although Europeans might be inclined to ask whether this was not a special syndrome of modelling: a ridiculous identification with American ideology. Yet, this behaviour would, according to Leaver, not differ structurally from the reactions elicited in the European allies (but adapted to their geopolitical situation) after each shift in American foreign policy.

Usually nationalist discourses tend to exaggerate the few historical episodes when the country went it alone against a hostile environment, but what is the deeper meaning of a myth of dependency the like of that the Australians seem to have cherished? What question about identity did or does it answer? Leaver suggests that this narrative served in the first place to neutralize the political effect of domestic economic problems by declaring them foreign policy problems. In this way they could be attributed to uncontrollable external factors such as the whims of faithless allies. Both the British loyalty to the European Community and American competition in traditional Australian markets were perceived as treachery. This theme fitted very well with the political strategy of the Labour Party. A second reason given by Leaver for the Australian 'myth of dependency' is simply that the entire Australian political culture and representation of the outside world has not yet succeeded in casting aside two centuries of imperial conventions.

THE NEW REALISM

The loosening of ties with Britain initially evoked isolation, which expressed itself particularly in economic protectionism and in the ideal of white Australia. The state of mind underlying this national condition has been summarized in labels such as 'Benthamite society'[22] or 'Frightened country'[23]. The first term refers to the philosophy of the English philosopher Jeremy Bentham (1748–1832), who interpreted social order on the basis of individual material interest, legalism and positivism. This view rejects all speculative and non-utilitarian thought and action. The phrase 'frightened country' denotes that visions of the outside world are excessively preoccupied with scenarios of danger. The utilitarian politics that were based on such principles have collapsed under their own contradictions. An extremely narrow conception of self-interest like that embodied in protectionism works in the current world order against, rather than in favour of, the national interest. The paranoid concentration on national security is likewise counterproductive.

It can be expected that a pragmatic people will revise a policy if it does not produce the desired results. This indeed has happened in Australia in the 1980s. The co-operation with countries in the region is the spearhead of the new foreign policy. However, many critical observers believe that the visions of the outside world have not evolved to a sufficient degree, or that they are not quite synchronous with international trends. The switch from protection to a complete liberalization of trade was not difficult to accomplish within the Australian pragmatic frame of mind. Initially some sectors would be harmed (industry) and others would take advantage of it (farming, mines), but in the long run the prospect is of an improvement of overall economic performance. At the very moment when Australia was fully committed to worldwide liberalization, a 'new protectionism' or 'economic nationalism' seemed to emerge in countries that formerly championed free trade – such as the United States and countries of Europe. Is Australia 'pursuing the correct strategies but in the wrong historical age'?[24] The cause of the new problems seems to be the anxious commitment to multilateralism of countries like Australia (and Canada) in fields requiring international co-operation, such as trade, security and the environment.[25] Why is Australia so committed to multilateralism in international arrangements? The answer may be found in an immature sense of national identity which needs to refer all questions of right or wrong to universal values. Multilateralism supposes universal values to be successful. In the European Union, which is often blamed for its new protectionism, some domains deemed socially valuable are shielded from market influences.[26] Such values do not have to completely disrupt a liberalization of trade but they suppose negotiations in which the parties are very well aware of their own social priorities and values.

'Locating' Australia in the world is not the same as either universalism or isolationism. It requires a self-definition which is difficult to make for countries that have always used the outside world as an excuse to escape such choices. Nor does the economy give definite answers on social priorities. In this way the attempt to end self-absorption by opening up the country to the world evokes the very question that it had hoped to escape. Who and what are we? In what direction are we heading? Wandering in circles . . .

THE EURASIAN DILEMMA

(*Russia*)

AN AMBIGUOUS FUTURE

Jack Matlock, US ambassador in the Soviet Union during the crucial years of the dissolving Soviet empire, has recently put down his experiences and expectations in an elaborate attempt at writing current Russian history. The message is that neither a return to the Soviet system of social and economic control, nor the reassembled Soviet empire is a conceivable future development, the first on account of the bad performance of the system compared with market economies, the second because such an imperialist development would stand in the way of the healthy economy that would be required to maintain imperial power.[1]

It is not difficult to broadly agree with this analysis, mainly because it specifies what will probably not happen. However, it still leaves open an uncomfortable range of options as to the direction of political and social change in Russia in the near future. Does it presage a Russian nation moving towards the Western type of democracy (the ultimate fate, according to Fukuyama, of all current dictatorships and totalitarian states)? And does it presume that, possibly after a transitional period of emotional attachment to neo-communist or nationalist ideals, Russian relations with the world outside will approximate the pragmatic relations that exist between Western democracies? Before entering into a discussion of the historical and geographical conditions shaping Russian world views, we must, first, recognize that countries' foreign policies are not necessarily in harmony with what the outsider defines as long-term interests (as even such 'pragmatic' players of the liberal world game as the United States and Britain have shown) and – second, that an emotional reaction to the new conditions, even for a transitional period, may turn out to be very troublesome to the world, certainly to Russia's neighbours.

In terms of historical (pre-Soviet) heritage, few countries would be more qualified to bear the title 'Russia's antipode' than the United States. The Russian empire was built on a long peasant tradition tying man to a fixed territory and always evoking strong feelings of insecurity and xenophobia; in spite of its avowed isolationism, the United States has a tradition of looking across frontiers and viewing its relations with the world as an expanding system which advances towards economic integration rather than partitioning of the world. Russia has known ethnic fragmentation and cultural fissures since the church schism (*raskol*) in 1666 and since Peter the Great's modernization drive divided society into a progressive elite and a conservative mass; the United States has a tradition of horizontal and vertical social mobility and denial of class. The question is, of course, how much Russian attitudes and outlooks have been transformed by the twentieth-century communist revolution and by events in the few years since the democratic revolution. The Bolshevik Revolution at one stroke upgraded social norms and values to something resembling American ideas of social equality with the exception, of course, of Marxist ideology. At least it (formally) abolished classes, introduced a cosmopolitan and missionary outlook, and attached a positive value to change – values and structures alien to the previous peasant society. But did they really catch on?

It is still difficult to tell what the meaning of the communist period will turn out to be for Russia's particular development in the coming decades. There are three logical possibilities regarding the impact of communism. First, it may have been an intermezzo, now followed by an urge to take up the threads broken in 1917 (model A–B–A'–).[2] Second, the communist regime may have laid the root for unintended consequences that were not intrinsic to the pre-revolutionary situation. By giving national differences a legal and administrative role, the communists may have hastened the disintegration of the empire (model A–B–C–). Third, the communist system may simply return in a more versatile shape (model A–B–B'–). As societies are complex phenomena, such models will probably never be corroborated entirely by events. They are more likely to have some value in classifying various features of the current social transformation.

Public opinion gives little hold on the current formation of leading ideas in Russian society. There appear to be deep divisions and uncertainties about the desirable course of affairs. A study of the structure of political beliefs, using research material gathered in the period from April to May 1992, revealed how strong doubts concerning economic and political reform still were among ordinary Russians.[3] The average score on these reform aims was about 5 (for economic reform) or 6 (for political reform) on a scale from 0 to 10 with 10 being complete agreement. A substantial difference is revealed between the answers from mass and elite respondents. The political elite more positively embraces ideas of political and economic reform (scoring on average one point higher) than does the public. Democratic principles, however, are more equally and positively appraised by both groups (with scores well over 7). This also goes for nationalistic attitudes which often prove to endanger democratic practices! The most intriguing result of this comparison of mass and elite belief systems was the low level of what is called 'belief constraint' among the Russian elite. Belief constraint refers to the internal cohesion of a set of statements about the world, as measured by comparing all group members. Technically it means correlation; a positive score on one item is time and again coupled with a positive (or negative) score on another item or the reverse. Results form Western European research reveal a much higher level of constraint in elite beliefs than in those of the mass, even higher than in those of the politically more involved citizens. Elites have to be accountable, they are accustomed to explaining and defending their opinions to political opponents, constituency groups, the mass media, and so on. They do not just give the answer that suits the mood of the day. The belief constraint of the Russian elite, however, turns out to be scarcely higher than the figure found for the group of most involved Russian citizens. This may be a comparatively good performance for the Russian citizen but it is a bad one for the elite. The authors suggest that the results indicate the lack of information-processing opportunities and political institutions that would normally promote consistency of issue preferences among the elite. However, one might also put forward the suggestion that the comparative newness of these ideas is an equally sufficient, and more straightforward. explanation for the low level of belief constraint.

Although the conclusion on 'belief constraint' may be considered a telling result of the above-mentioned research, one may doubt whether this type of belief-system analysis can yield much sensible insight. The problem of post-communist Russia is the utter instability of political language compared to distinctions in Western political discourse. For example, politicians are in the habit of projecting their own identity against a phantom Other by choosing the designations with the most battering impact in any given circumstances. As Michael Urban has illustrated from an interview with a leader of the 'Civic Union' coalition, even supporters of 'reform' may be accused of having a 'Bolshevik mentality':

> Zharikhin: Yes, it is a classic variant of the ends justifying the means, destroying everything for the sake of creating a new social order. a purely Bolshevik thing which is not limited to communist ideology. Ideology itself is not the main thing. In order to achieve their aims, they use liberal ideology but remain Bolsheviks.[4]

A little further on during the conversation Zharikhin identifies communists as fascists. In another situation a protracted discussion is triggered about the name of a new electoral bloc. Each proposed term is successively rejected: 'democratic' because it has a bad connotation (like 'privatization'), 'united democratic left' because in the West it stands for parties that are insignificant, 'socialist' because it is not attractive either, and in the end even the word 'party' turns out to be suspicious. Ideas seem constantly to be crippled by the weight of references to points elsewhere – the Russian past, the West – leading to rejection of what may be of significance for the solution of practical problems. Clearly, as Urban states, 'politics in postcommunist societies is in large measure a politics of identity'. We might as well say 'a politics of paranoia'.

What about foreign relations? An indeterminacy, something like a lack of constraint in the beliefs of one person, may be recognized in Gorbachev's shifting definitions of the identity of the Soviet Union. In his book *Perestroika*, Gorbachev takes offence at those who want to exclude the Soviet Union from Europe and calls Europe 'our common house'. This is underpinned by references to the common heritage of the Renaissance and the Enlightenment, religion (Christianity) and the struggle against Hitler's fascism. But Gorbachev appears to have more common heritages in mind:

> To ascertain how many common houses apart from Europe, the Soviet Union really has, it might be useful to listen carefully to what the peripatetic Soviet leader says about every outpost of the Empire's frontier. In July 1986, Gorbachev was waxing lyrical in Vladivostok about the Pacific frontier – our common house; a few months later, he was in India stating something similar about the Indo-Soviet friendship with regard to the Indian Ocean region; almost a year later, in October 1987, Gorbachev went to Murmansk, where he appealed to the Scandinavians and Canadians mentioning the Arctic – our common house. The problem seems to be, as Martin Walker pungently captured it in his article on the importance of geography in Russian history, that 'the Soviet Union is so damned big that it has too many common homes for the comfort of its many neighbors'.[5]

Even the geographical (geopolitical) setting of the Russian state seems to pose problems of identity. These doubts were repressed at the time of the Soviet Union, when the empire was kept together by military might and an inward-looking orientation. The terms 'East' and 'West' at that time only referred to the ideological struggle of the Cold War. But since the take-off of *perestroika* they have gradually come to denote a geopolitical confrontation between Western Europe and the Far East in which Russia becomes either a part or a mediator. Only those who continue to believe in Russia's 'great power' status have no problem of choice or priority. As Leonid Abalkin has stated:

> The geopolitical position of Russia makes a multi-directional orientation of its foreign policy and its inclusion in all enclaves of world society an objective necessity. Any attempt to put at the head of the list its relations with one side or group of countries is contrary to its state and national interests . . . It is true, the question of the geopolitical priorities of many countries is legitimate, but nor for Russia as a great world power.[6]

Great power status or not, the ordinary Russian citizen has in large measure lost belief in grand goals. Results from citizen surveys in 1989 and 1994 suggest that the awareness of a common national goal faded even more during those five years. In 1989, in answering the question what 'the main tasks facing the country' were, the option 'material prosperity' was chosen by 40 per cent of the respondents. The highest scoring option in a more or less comparable question in 1994, only represented 20 per cent of the respondents ('legality and order').[7] The breakup of the Soviet Union was in 1994 assigned third place in a rank ordering of 'most important events of the twentieth century'. The top score was for the 'Great Patriotic War' (the Second World War) and 'October 1917' was accorded second place. Very few respondents approved of the breakup of the Soviet Union.

All the facts from opinion research suggest that Russian society currently lacks positive ideas capable of attracting a large part of the population and that the public may be responsive to the programme of any counter-elite able to articulate new aims in a more consistent manner. Does this mean that Russia's foreign relations and visions of the world are completely unpredictable? Not entirely. The answer depends on the potential strength of what one might call the geopolitical reflex. Just as the human motor intelligence takes care of bodily reactions, even if the brain is confused, a geopolitical reflex would cause co-ordinated reactions to the world outside even in the absence of a stable political programme. Changing configurations in the world around would trigger such responses, and possibly domestic events would do the same. A geopolitical reflex might even be triggered by a hallucination like the 'phantom limb' pain in people with amputated limbs.

For those who dislike organic analogies the problem may be put in different words. In a situation where thoughts on the future course of society are contradictory and casual, the concepts most capable of mobilizing different people and interests belong to the nationalist repertory. Nationalist visions emphasize the common descent and blessings bestowed on a 'chosen' people. They usually involve symbolic (holy) places and geographical destinies. Authoritarian, communist and even liberal ideas about society can accommodate – at least partly – a common set of nationalist ideas. The idea of national security unifies classes and interests and it is possible, particularly in Russia, to assume a profound sense of physical insecurity towards the world outside. Leaving aside the old peasant psychology, this feeling was reinforced by a history of violent invasions from Napoleon to Hitler, by the asymmetrical geostrategic world order during the Cold War, and by antagonisms with other cultures (Islam) within the state and in the 'near abroad' (countries of the former Soviet Union).

BETWEEN WARM SEAS AND ARCTIC SALVATION

National identity, the question of what and who is Russian, has never found a straightforward answer. That more than one early variant of national consciousness has been distinguished is symptomatic for the fissures in Russia's socio-political structure. One interpretation of identity focused on the political ideology of the state, another was based on popular ethno-religious identity, and a third originated among the intelligentsia.[8]

The earliest Russian political consciousness was based simply on the fact of being a subject of the tsar and living within the territory of the Russian state. The Orthodox faith was originally also included in this conception of the state but in the course of time the tsars downplayed this element. The Russian empire contained other cultures and religions than the Christian faith and the state had to remain neutral between such differences. The missionary activities of the Church were even restricted in newly conquered Tatar territories in the sixteenth and seventeenth centuries. The secularizing activities of the tsars in their attempts to emulate Western European developments and their hostility towards manifestations of Russian nationalism, fit this pattern. Currently, some Russian intellectuals, such as philosopher Jurij Borodaj, still long for a resurgence of the imperial system:

The unity of orthodox, slavic nations was kept in sway by the fact that there was one rightful Sovereign of Russia. And it was without importance where the capital was – in Minsk, Kiev or Constantinople, as Dostoevskij dreamed, or in Petersburg. In general, this had no significance. What was important was the core of the union – a single, rightful sovereign standing at the head of the family of related autonomous peoples, ethnic units closely knit in spirit as well as sub-ethnic groups ... The Soviet Union was totalitarian and the Russian empire was not totalitarian. It was a union of vital nations, communities, confessions, corporations; it was founded on the principle of many strata.[9]

This ideal could not prevent the peasant mass developing a popular identification with 'Holy Russia', the Russia of the Orthodox faith, rather than with the ruling dynasty. It even implied a hostile attitude toward the tsars. Peter the Great was called an infidel foreigner and servant of the Anti-Christ. This popular belief sometimes reverted to a particular geopolitical vision: the theory of the Third Rome. The fall of Constantinople (the second Rome) to the Ottoman Turks in 1453 had left Moscow as the single remaining spiritual centre of a Christian empire. During the reign of Basil III (1505–33), the idea of a God-given task of preserving the faith – in the sense that Moscow was also the *last* Rome – was more or less sanctified, but later successors to the throne distanced themselves from this idea.

Finally, the intelligentsia tried various solutions to the problem of Russian identity without referring either to the autocratic regime or to Orthodox religion. This movement often sought refuge in Pan-slavism or in Russian cultural chauvinism. The expeditions of geographers to the Amur river in the 1840s and 1850s 'to bring civilisation to Asia' can be seen in this light. It is not surprising to learn that Tsar Nicholas I was opposed to such expeditions.[10] Even these approaches to the problem of national identity were not able to launch a modern nationalist movement, of the type that supported democratic and industrial impulses in other European countries. 'Bringing civilization to others' primarily served as a means of self-distinction.

The impotence of this partitioned nationalism to create an unified mass-movement inevitably ended in the political earthquake of 1917. But since the Bolshevik Revolution did not elaborate on any of these nationalistic themes, one might assume that its collapse cleared the way for the resumption of old, unfinished business (A–B–A'–). And new heroes of the nationalist cause have indeed announced themselves without delay.

The best-known and most steadfast champion of the nationalist cause is the writer Alexander Solzhenitsyn. He aired his opinion at the height of the post-Stalinist repression, during his exile in the United States and after his return to Russia. Here someone is speaking who is averse to fashionable and opportunistic rhetoric, who shows no mercy to friend or foe. Contrary to Gorbachev, Solzhenitsyn sees the Renaissance and Enlightenment not as a common home to be proud of, but as the onset of the decline of the West. The West turned its back on spiritual values by emphasizing material consumption and competition, it promises freedom but offers manipulation and false thought. Instead of yearning for power, wealth and expansion, one should turn to creating domestic order and spiritual harmony. 'Should we be struggling for warm seas far away, or ensuring that warmth rather than enmity flows between citizens?' is the choice Solzhenitsyn presents. With that perspective he appeals to the Russian longing for security and also links up with old ideas on Moscow as the Third Rome and the Holy mission of Russia.

The geographical destiny of the Russian people, according to Solzhenitsyn, is the emptiness of north-east Siberia. Apart from the salutary effect of a 'frontier-experience' on the Russian soul, this internal colonization plan was an attempt to cope with the future Chinese danger. It is striking how since the late 1960s, Kremlin rulers and dissidents like Andrej Amalrik and Solzhenitsyn were united in their fear of the Chinese. The dissidents, of course, saw war with China as a cunning way to challenge the simplistic communist interpretations of the struggle between good and evil. They did not succeed in changing the communist system, which could assimilate this new fear also. Ultimately China may indeed prove to have contributed to the fall of the Russian empire, not by war, as we know, but because China demonstrated that relaxation of communist economic control was immediately followed by an economic boom, whereas the Soviet Union muddled on under aggravating economic conditions.

Solzhenitsyn's ideas about introspection and self-realization may satisfy the desire for a special role for Russia in the world, something which cannot be identified with either communism or capitalism, but his austere approach is not likely to attract much following.

A completely different nationalistic discourse is entertained by Jirinovsky. Jirinovsky, leader of the Liberal Democratic Party, does not accept the dissolution of the Russian empire like so many others. He more or less wants to restore the pre-revolutionary situation, including a German expansion into Eastern Europe (East Prussia!).[11] Poland would be reduced, Slovakia, Belarus, Ukraine and the Baltic republics annexed, and the Russian sphere of influence in the Balkans restored. With a 'last dash to the south' Jirinovsky wants to give Russia access to warm waters, the Indian Ocean and the Mediterranean. Equilibrium in the world will be restored if four superpowers with their own spheres of influence are created: Japan dominating East Asia and Australia, Russia dominating South Asia, Western Europe with its former African colonies, and the American hemisphere with the United States as dominant power.

Jirinovsky looks nostalgically to the balance of power in Europe before the First World War. That is why even he allows Germany, the state which has caused Russia's biggest war trauma, to expand. Possibly, Jirinovsky hopes that a stronger Germany will compel the United States to withdraw from Europe. A unified Europe with Germany as core state is, even if it is militarily unified, more bearable than the continued presence of the United States near Russia's borders. Such ideas are reinforced by the impression that the multipolar world, forecast at the end of the Cold War, does not quite seem to be materializing. As many small crises and diplomatic initiatives have shown since 1989 – in the Gulf, Somalia, Bosnia and Israel – the United Nations is simply the instrument serving the United States to create its own world order. That is, at least, how nationalist Russians experience current world events.

Not every Russian considers the United States as the country's nastiest antagonist. The younger generation assimilates US culture, and most people are more concerned with 'alien' elements either in Russia itself (Jews![12]) or at its southern borders (Muslims) than with US actions far away. This means that certain features of Jirinovsky's scheme, and particularly its aggressive aspects, will not be widely appreciated.

More or less assimilated in these extreme options are ideas propagated in Russian émigré circles after the Bolshevik Revolution. Since the recent fall of the communist system, some of them have won a new freshness because they in fact were a pure Russian alternative to the communist revolution. The doctrine of 'Eurasianism' rejects the internationalist (Western) orientation of both tsars and Bolsheviks; it rejects linear development in history as well as the value of mass movements.[13] Its individualized and 'post-modern' features may very well satisfy some current demands for intellectual guidance. The main originators, Prince Nikolay Trubetskoy and Pyotr Savitsky (a geographer) tried in the 1920s and 1930s to give a theoretical underpinning to the identity of Russia as a non-European state. They downplayed the significance of the Urals by drawing attention to the parallel zones of humidity and the corresponding vegetation zones which characterize Siberia and European Russia but not Western Europe. They tried to derive the character of Russian society from the 'perennial struggle between steppe and taiga' and used the diffusion of Orthodox Christianity as the main indicator for the delimitation of Russian territory. They thought that Russia as a non-European and non-Asiatic country had a calling to protect Asia against Eurocentric approaches and colonialism (an attitude tried out again by Soviet regimes after the Second World War).[14]

None of these visions is limited to the territory which is currently called the Russian Federation. Even Solzhenitsyn favoured a Great Russian solution. The common denominator of all nationalist discourse seems to be the attribution of a non-Western identity to Russia and the avowal of the unity of the Eurasian territory, an area which more or less corresponds with the pre-revolutionary empire, but which does not necessarily correspond with a unitary state. The implications of this geopolitical vision vary a lot among different opinion-leaders with a nationalistic outlook.[15] At one extreme of the continuum we find those who sympathize with democratic reform but who nonetheless think that Russia still needs to exercise its authority in the 'near abroad'. At the other we find those who advocate the downright restoration of the former Soviet Union or who even think about more awesome imperial versions.

A RUSSIAN 'MONROE DOCTRINE'

If the geopolitical reflex is so clearly geared to the territory of the former empire(s) and not to the current Russian Federation (RSFSR), why then did the Soviet Union fall apart? One might think of several reasons for this development which still rouses mixed feelings in many Russian minds.

A first answer is that the institutional structure of the Soviet Union, with its built-in acknowledgement of ethnic territorial separation, paralysed such a reflex.[16] In fact the fifteen Union republics, each named after a particular national group, had been assigned the right to secede from the Union. Western Sovietologists used to emphasize the purely formal and theoretical nature of this provision. Although this was a realistic understanding of conditions under stable communist rule, it suppressed the awareness that this very structure could also facilitate a Soviet breakup under particular circumstances (model A–B–C–). Such unexpected circumstances occurred when communist ideology was revoked as a political line of action in the Soviet Union. For one thing the Russian Federation was similar to the other republics in demanding the dissolution of the Union because the Soviet Union was associated with communist power. For another thing the Russians were the only ethnic group to lose power by this operation. As Rogers Brubaker aptly formulates the dilemma:

> The jurisdictional struggles of the RSFSR against the Soviet centre were [. . .] two-sided, oriented on the one hand to *weakening* the centre and distributing its powers to the national republics, and on the other hand to *capturing* the centre and taking over its powers.[17]

But the latter strategy was an unknown and unsure line of action. It still is unclear what the wielding of power by the new republics, or even what their 'statehood', will mean. This may be one reason for the apparent absence of what I have called a geopolitical reflex, or in any case for the absence of a tangible result. This could not materialize because the old institutional framework was not intact any more and because the new 'body' was strange. Russians may certainly have felt a more profound tie with the territory of the Union than other nationalities, but they could not articulate that feeling because national groups did not have a voice other than as republics. That was the legacy of the old system which now took a surprising and uncomfortable turn.

Another possible answer is simply that a geopolitical reflex is not always active. As Hannes Adomeit suggests, there was an ebb and flow of two dominant paradigms in the former Soviet Union and its successor state Russia: the 'Ideological and Imperial' and the 'New Thinking'.[18] The New Thinking emerged in the mid-1980s and lasted until the end of 1992. The main ingredients of this New Thinking were that military power, geopolitical expansionism and empire-building are outdated forms of international conduct; that status and power in international affairs are determined by economic efficiency and human resources; and that interests have to be promoted through multilateral approaches and participation in international institutions. Since the end of the 1970s it had become clear that military power had not resulted in enhanced power and prestige for the Soviet Union in the international arena. Its economic model was not deemed worth imitating by Third World countries and even its military power did not bring about the desired state of affairs abroad (as evidenced by the resistance to the occupation of Afghanistan, mass resistance in Poland, the stationing of intermediate range cruise missiles in Western Europe). The New Thinking promised a revitalized Russia irrespective of the status of the territories within the borders of the Soviet empire. The example of Germany and Japan taught that free-market conditions enhance economic strength and international status of a country far beyond the limits that seem to be set by its territory. Russia, after all, would never be reduced to a really small territory, it would remain the dominant member by far in any conceivable Commonwealth of Independent States (CIS).

This suggests a third reason for the apparent absence of a geopolitical reflex during the dissolution of the Soviet empire: even the New Thinking could assume the preservation of a special relationship between the Russian mother country and the parts of the former Union. There are indeed compelling reasons for maintaining mutual relations. These have to do with the mutual dependency on mineral resources (coal, oil, gas), the common military infrastructure, and the insufficient resources for maintaining national security in some of the new republics. The security problem is the least symmetrical dependency in the new Commonwealth of Independent States. It is, in any case, particularly emphasized by the Russian leaders who currently follow a military doctrine (adopted on 2 November 1993) that the deployment of Russian troops in CIS-member territories is allowed for 'peace-keeping' functions, for protecting Russian minorities and to meet a possible attack on Russia.[19] This concept is often designated the Russian 'Monroe Doctrine'.

Since the end of 1992 the 'Ideological and Imperial' paradigm has come to the fore – again. Owing to the fact that economic growth failed to materialize, to the continuing harsh living conditions of most citizens, and the feeling that Foreign Minister Kozyrev had conceded too much to the West, the demand for reform and even more personal sacrifice in order to earn an uncertain paradise in the future became less convincing. A new assertiveness in foreign relations will certainly satisfy those whose feelings are injured by the confluence of individual impoverishment and national humiliation, but it will hardly affect living conditions themselves. It will probably not even aim at restoring the old system with its relative international isolation. What happened was that the discourse around universal values which replaced communist ideology had to leave the field to a new emphasis on Russian national interests. One came to realize that the new international system did not automatically imply the coincidence of Russian interests with those of the West. The idea of opposite interests also implies a heightened geopolitical awareness, not necessarily the old geopolitics dominated by questions of military security and power, but an idea about economically belonging to a certain area and having diverging relations with other blocs or countries.[20] A stronger pressure on the former parts of the Union, the CIS, would be the most immediate consequence of the new prevailing 'Ideological and Imperial' paradigm.

The relations with the second-largest former member of the Soviet Union, Ukraine, are the most important and complicated in this respect. Ukraine is hardly foreign territory for Russians. In the eastern part, the Donbass region, 44 per cent of the inhabitants are Russian and an even higher percentage are native Russian speakers. Moreover Ukraine is the cradle of Russian culture, as Kosovo retains the memory of the Serbian past. As an old proverb says: 'Moscow is the heart of Russia, St Petersburg its head, but Kiev its mother.' Already in the ninth century, merchants from the northern Russian territories descended along the rivers Don, Volga and Dnieper until they finally emerged at Constantinople in 860. In the next century they founded the princedom of Kiev and converted to Christianity around 1000. They were the founding fathers of the Russian nation and, finally, the sole heirs to the Byzantine branch of Christianity. The importance of Ukraine for Russia is both symbolic and strategic. In terms of population and natural resources, Ukraine would add substantially to Russia's power. Zbigniew Brzezinski stated: 'It cannot be stressed strongly enough that without Ukraine, Russia ceases to be an empire, but with Ukraine suborned and then subordinated, Russia automatically becomes an empire.'[21] The future course of Russian politics is not inevitably aggressive but if it does turn out to be, the odds are that war with Ukraine will break out.

Another important geopolitical shift has resulted from the independence of the five Central Asian states: Kazakhstan, Kirgizstan, Tadzhikistan, Turkmenistan and Uzbekistan. The termination of Russian rule in Central Asia and the Caucasus means that Russia is no longer a 'Middle Eastern power'.[22] Apart from this 'problem', 9 million ethnic Russians are living in the area (more than 6 million in Kazakhstan alone) and Russian troops are still stationed in each state. However, Moscow did not resort to destabilizing attempts or to exerting hostile military or economic pressure on these states (or in Ukraine). The Russian claim that its military presence is required for maintaining stability is acknowledged by the Central Asian authorities, although Russian troops were placed under their jurisdiction in three states (Kazakhstan, Kirgizstan and Uzbekistan).

Does the loss of the 'Middle Eastern connection' touch the same nerves as the separation of Ukraine? In spite of Jirinovsky's rhetoric, access to the Indian Ocean or Persian Gulf never had strategic priority in Russian history. Military mock attacks aimed at the Indian subcontinent during the nineteenth century were primarily meant as a diversion, a means to alleviate British military pressure in the Near East. 'Russian strategists never accorded Central Asia the same priority as the potential war theatres in Europe or the Far East,' Milan Hauner states.[23] The military build-up along the southern border since the 1960s was dwarfed by the concentration of troops and military equipment in the Western and Far Eastern 'theatres of military activities' (TVDs). Nonetheless, a communist coup in Afghanistan was supported in 1978 and a year later the Soviet Union invaded that country with a large military force. The diplomatic flop in Egypt, which revoked its treaty of friendship with the Soviet Union in 1976, also probably inspired a search for alternative access to the southern seas. Why this rising interest in the southern geostrategic realm in recent times? One answer is the need to provide naval bases for the fast-expanding war fleet. Another is fear of disruption to the vulnerable trans-Siberian connection (most probably by a Chinese attack) and the possibilities, in such circumstances, of reaching the Pacific coast by sea.

Currently, the fear of a war with China has somewhat subsided. There is even the feeling that stability in Central Asia, which is possibly threatened by future Muslim fundamentalism, is a common interest of both Russia and China. This explains why the loss of the Central Asian republics was in itself not the most sensitive trigger for a geopolitical reflex. However, changes in spheres of influence of other powers (Iran, Turkey, China or the United States) will be closely watched by Russia and the possibility of a military answer to external interference is unequivocally asserted. The term 'Russian Monroe Doctrine' appropriately describes the geopolitical sensitivity of this region in the Russian perception.

The Caucasian region, however, offers a quite different perspective. Here Russian interference entails much more than the avowed reaction to imaginary security threats. Russia's alleged complicity in the overthrow of Azerbaijan's President Elchibey (1993) after this leader ventured to build a pipeline in collaboration with Turkey, Russian support for the secession of Abkhazia from Georgia, and the dramatic repression of the Chechnyan revolt (although a legitimate part of the RSFSR) suggest that the most sensitive geopolitical nerves are indeed to be found in this region.[24] One of the underlying motives for Russian interference is the traditional rivalry with Turkey and the fear that the Caucasian republics might back out of the CIS economic system, even transforming themselves into forces competing with the RSFSR in foreign markets (oil). Turkey's threat is not limited to the new international sphere of the CIS countries. It has also established direct contacts with Turkish-speaking and Muslim minorities within the RSFSR. The Balkans are Russia's traditional destination if it wants to outflank Turkey. Therefore the Russian idea of building, with the co-operation of Bulgaria and Greece, a pipeline from the Black Sea to the Mediterranean is a manoeuvre that fits wonderfully into time-honoured tradition. However, the loss of spatial contiguity with this part of Europe complicates Russian manoeuvrability.

THE EURASIAN IMPERATIVE

After the respectable period of three centuries, a common state including three Slavic peoples, the Russians, Ukrainians and Belarussians, was overturned. This, even more than the loss of the three Baltic states, meant that Russia (as successor to the former empire) was bereft of territories that had formed bridges with the rest of Europe. Ukraine was not only the symbol of old Kievian Russia with its Orthodox roots, but also the place where more secular traditions integrated with West European culture flourished.[25] This reminds us of the deep crisis of identity which was unleashed by the dissolution of the Soviet Union.

The shift to the East (not in a geographical but rather in a perceptual sense[26]) has heightened Russian awareness of Asia and revitalized the old

'Eurasian' body of thought. This means that keeping the Eastern territory with its Pacific Ocean frontier became a more pervasive element of the geopolitical vision. But currently this does not necessarily imply the establishment of a fortified border zone and the aim of geographical isolation. Even imperialists who continue to see the CIS as an empire acknowledge that the structure of the world around has changed. They are aware that the economic viability of a geographical area, particularly one so far separated from the (European) core area, cannot be warranted by internal relations alone. The maintenance of Eurasian power seems to require an Asian connection, which is under-pinned by the idea of a contrast between two 'modes of production', the conservative and individualist West against the dynamic, collectivist East. The Eurasian mind would fit in more easily with the latter, turning a former drawback (like the collectivism and the low level of development) into an advantage. Such wishful thinking resembles reactions in national peripheries such as Corsica, Scotland or Sicily in advance of the Single European Act. Here traditional structures were sometimes praised, even by social analysts, as a real blessing in the post-modern world. In the same vein, Borodaj praises the Eurasian disposition for the future of Russia:

> The irony of history is, that those preserves in the Far East not yet poisoned by the all-destructive 'civilizing function' and still defending their autonomy, the creative potential of the half-primitive craftsman, who has preserved the communal-family habits of social life, turn out to be far better adjusted to the newest technologies. And this craftsman is not in the West but in the Orient. The decline of Europe is becoming a reality before our eyes.[27]

However, this non-European identity is not easily adopted by many other intellectuals who have become accustomed to such a long tradition of cultural identification with Europe. Their hopes may depend on the interpretation of Eurasia as a unique synthesis of European and Asian impulses. For that option they hardly have any choice but to endorse attempts to maintain Russia's status as a great power able to occupy an autonomous intermediate position between Europe and Asia. Only that may both save European identity and admit a new edge of Asian magic to Russia, at least from an ideological point of view. Whether great-power status can be combined with new economic *élan* and whether the Asian touch will have any significant contribution to make remains questionable.

Thus, a Russian geopolitical reflex seems to be aroused by the changing international reality, the emergence of Asia and Europe as separate economic entities, and Russia's economic dependency on the global economy, as well as by Russia's internal eastward shift. The new geopolitical thinking is not only the avocation of a conservative or military elite but derives its strength from the fact that it accommodates rather divergent interests and ideals, even those of the liberal intellectual (although perhaps willy-nilly). The geopolitical reflex is recognizable in a heightened vigilance against internal separatism in the RSFSR (Chechnya) and against centrifugal economic tendencies in the countries of the CIS periphery. The perception of an eastward shift means that the geopolitical reflex is not simply geared to restoration of the former Union (in which case reintegration of Ukraine and Belarus would have got the highest priority), at least if we assume that the perception of an eastward shift also means the acceptance of its consequences. After all, surprising changes on the 'western front' can never be excluded. At the moment I am inclined to interpret current policies and aims as attempts not to reproduce Soviet-Russian, but to reconstruct Russian, identity under changed circumstances. In that respect the organic analogy fails because it would accord predictable power to a fixed primeval territory (like that of the Romanov empire). Nothing can be excluded from change in the life of a nation but it is also true that many things cannot change at the same time. To expect both a territorial reshuffling and the end of a dominant Russian role in the 'near abroad' (that is, in an Asian sphere) is undoubtedly too much.

THE EMPIRE OF REVENGE

(*Serbia*)[1]

HISTORICAL REVERBERATIONS

'My dear friend,' writes a Serb soldier, 'appalling things are going on. I am terrified of them . . . . I dare not tell you more, but I may say Ljuma . . . no longer exists. There is nothing but corpses and ashes.' A Franciscan, who went there, told me of the bodies of the poor little bayoneted babies. 'There are villages of 100, 150, 200 houses where there is literally not a single man. We collect them in parties of forty to fifty and bayonet them to the last one'. The paper says it cannot publish the details, 'they are too heart-rending.'

Nothing could make the luckless refugees believe that the Great Powers had really given them to the Serbs. They asked piteously when the UN was coming to drive the Serbs out. And still the Powers did nothing. . . .

It was feared the Serbs would descend on Elbasan, and I carried away a whole mule-load of valuables to save them from being pillaged . . . I learnt from the refugees that twenty-six villages had been wholly or partially burnt and pillaged by the Serbs. Few of the refugees had any weapons. I reported all this in vain in Scutari. Not a Power would move. The Serbs, grown impudent, then entered strictly Albanian territory in defiance of the International forces . . .

At the Foreign Office I begged protection for the Balkan Moslems, who were being barbarously exterminated . . . . Those with whom I spoke admitted that the consular reports from Uskub and Monastir were very bad but that it was not advisable to publish them. In truth we were hopelessly tied to Russia and could say nothing about her pet lambs . . .

It was also made clear by the Mazowiecky Commission that the accusation that the atrocities were planned and carried out by the Serb 'Black Hand' society were true . . . All was part of a prearranged Great-Serbian plan. 'The Serbs,' I overheard two Montenegrins say in the inn at Rijeka, 'are right. They put these gentry (non-Serb population) to the sword as they go, and clean the land.' As the Black Hand was a 'government within the government,' and not official, Belgrade could always pretend to be ignorant of its doings.

A superficial reading of these quotations would easily suggest that they are from a report by a war correspondent in former Yugoslavia in the 1990s. In fact they are observations dating from the eve of the First World War. The text describes the atrocities in the Balkan Wars (1912, 1913) but was much later included in a book of travel stories.[2] I have only changed the two most incongruous words: 'Prince' has become 'UN' and 'Carnegie Commission' 'Mazowiecky Commission'. The author of this text, Mary Edith Durham (1863–1944), was one of those fearless Victorian ladies who were attracted to the Orient (which included the Balkans at the time) as if by a magnet. Educated at art academy and specialized in the illustration of works of natural history (on reptiles), she first travelled to the territory of former Yugoslavia for health reasons. Her interest in nature soon gave way to the study of the area's customs and traditions, prominent among which are the many proposals by Serbian soldiers that idly hang around. In any case, politics is never far away in Edith Durham's travel tales. One of the journeys recounted in an earlier work, *Through the Lands of the Serb* (1904), carries her across the border of Montenegro into the Turkish-ruled 'Old Serbia', the town of Pristine, a journey full of dangers because the Turkish soldiers are known to be rapists. A Serbian woman tells her that she will be safe: '. . . the Turks will not dare touch you. They are afraid of your friends across the frontier and know you would be nobly avenged'.[3] Durham is baffled: 'a dark life where "safety" depends on power of revenge,' she muses.

Whoever recollects the more or less veiled threats of attacks in Washington and London made by the soldier Mladić and the politician Karadžić, must conclude that the 'power of revenge' is one of the more constant principles in Serbia's entanglements with the world outside. One has to be careful with generalizations based on the wisdom of hindsight, but the current susceptibility of some commentators to drawing a conclusion that may be interpreted as a statement on innate characteristics of one of the peoples of former Yugoslavia produces imbalances as well. This intellectual restraint is usually expressed in emphatic statements about the shallow roots of violence in former Yugoslavia and by a search for external causes or culprits. The Yale historian Ivo Banac also starts an interesting essay on the spiral of violence and its local causes with a rebuke to respectable newspapers for assuming 'the irascible hatred that governs the Balkan savages'.[4] The tradition of mutual hostility between ethnic groups is, according to Banac, no older than the end of the nineteenth century, at most. This 'at most' is already quite old compared with the vision of Michael Ignatieff who, in a review of a newly published book by his colleague Misha Glenny, suggests 1928 as a milestone, the year of the attack on Croatian parliamentarians in Belgrade.[5] Glenny himself, however, does not recoil from making such statements in his book as: 'The nationalist sinews in this region have been exercised and strengthened by centuries of violence and uncertainty.'[6] Other writers mainly focus on the inter-ethnic violence of the Second World War as a factor in the current war.

The problem is complicated because it is possible to understand the influence of the past in at least three ways: through a hereditary national character (the savages of the Balkans), through slowly changing institutional or social-geographical structures (an undeveloped civic culture, low degree of urbanization), and through the traumas and peak-experiences of the past, which under appropriate circumstances may reinforce feelings of hatred or affinity (like Great Serbia). There is little controversy between commentators about the relevance of the latter mechanism. And the comparatively long period of peaceful cohabitation of the Yugoslavian nationalities after the Second World War also casts doubts on the relevance of the first two of the three mechanisms mentioned above. But this does not mean that one should search for the causes of violence mainly in external reactions such as the ill-advised European recognition of Slovenia and, particularly, of Croatia. Tension between the different nationalities (republics) within the federal state of Yugoslavia had already been on the rise for a long time and would have led to disintegration, with or without the permission of the outside world. Even at the beginning of the First World War there was a large degree of unanimity in Serbia about the preferability of a strong Serbian state containing all Serbs, despite widespread propaganda for the Yugoslavian ('Illyrian') ideal.[7] Neither was a stable 'communist solution' of the nationalities problem discovered after the Second World War, even though this impression is easily gained, as a result of almost half a century of relative peace. There was a fragile balance that relied on the shock of the events of the war (in which at least half of the 1 million Yugoslav victims were victims of inter-ethnic violence), and on the prestige of Tito, particularly his skilful balancing act. The stupor induced by the Second World War lost its hold, not only in Yugoslavia but in many other countries in Europe, from the end of the 1960s. This was first shown in protests by the younger generation against authority and cultural repression in West European countries. In Eastern Europe it assumed the shape of increasing nationalism.

For some time the communist 'taboo' on any expression of national 'chauvinism' worked reasonably well. It was also endorsed by 'cosmopolitan' intellectuals such as the writer Dobrica Ćosić (soon to be President of Serbia) and the Croatian writer Slavenka Drakulić, both of whom preferred to call themselves Yugoslav rather than Serb or Croat. Such voices could count on an instinctive sympathy in Western intellectual circles, which partly explains the hesitant reactions in the rest of Europe to events in Yugoslavia. If clever Yugoslav men and women told us that one does not have to concede to nationalist aims, who could object? But intellectuals, and particularly writers, do not always reflect the range of sentiments in society. There was a strong undercurrent that experienced the emphasis on Yugoslavism particularly as cultural repression. Strikingly this feeling first manifested itself in a nation that could boast of neither a glorious past nor former independence as a state: the Slovenes. Already in 1965 the Slovene writer Josip Vidmar clashed with Ćosić in an article in which he denounced the idea of cultural assimilation as a process in which the Slovenes and Macedonians particularly would have to adapt to the rest.[8] In this article still another taboo was broken down: Vidmar asked for publication of the national accounts in order to provide a clear insight into the distribution of the federal burden over the different republics. An irritated Ćosić promptly accused Vidmar of narrow-mindedness and a lack of any feeling of responsibility for settling the delicate problem of inter-national (= ethnic) relations.

GREATER SERBIA

Distrust between the nationalities could no longer be disguised. In the Serbian community it burst out in the much talked-of Memorandum of the Serbian Academy of Sciences and Arts (SANU) in 1986. Now, it was Ćosić himself – an important contributor to the report – who expressed nationalist sentiments. The document established the existence of an 'anti-Serbian coalition' and even 'genocide' of Serbians by the Albanians in Kosovo. A whole gamut of fears emerged in Serbia, fears which Banac illustrated with cartoons by Milenko Mihajlović published in 1989: the Croats as Ustashas (fascist groups during the Second World War who murdered many Serbs), the imperialist aims of the Catholic Church and the Muslims, the Albanians as rapists, and so on. One of Mihajlović's cartoons (Figure 6) shows a little Serb boy who is almost being torn apart between a Roman Catholic bishop and a Muslim imam. The former exclaims 'Catholicize!' and plucks out the child's eyes (a reminiscence of Ustasha practices), while the Muslim, wielding a cut throat razor, shouts 'Circumcize!'. Such references to inter-ethnic violence were not confined to drawings and texts. In the streets provocative youngsters adopted the clothing, haircut and manners of the Chetniks (Serbian nationalist forces in the Second World War) or Ustashas (of Croatia).

Where nationalist sentiments prevail, the past is always important. It is the main source of arguments and frustration, however distorted and mythologized. There is basically no age limit to dates of possible significance and the frequent occurrence of events from the remote past in current political discourse is often surprising for people belonging to nations that derive their identity from seventeenth- or eighteenth-century trade capitalism or from the nineteenth-century wave of nationalism. The person who experiences five centuries of Turkish suppression as a gap in national history rapidly ends up in medieval Great Serbia, an empire that extended from a line between Belgrade and Dubrovnik south to the waters separating the Peloponnesus peninsula from the Greek mainland. On Kosovo Day (28 June), the Serbs traditionally commemorate the fall of the Serbian empire at the Battle of Kosovo in 1389 and the start of centuries of Turkish rule.

It seems peculiar to choose an absolute nadir in national history as the most significant date symbolizing national unity. The Serbs have a legend which turns this painful event into something glorious. Shortly before the battle, an angel appeared to the Serbian leader 'Tsar' Lazar (his status was upgraded in later Serbian tales). Lazar was presented a choice between victory, which would only yield an earthly kingdom, or defeat, which would ensure the Serbian people a place in heaven. This legend associates the Serbian national fate with 'martyrdom' but also provides the idea of a 'God chosen' people.

The Serbian areas in Bosnia and Croatia can apparently not be claimed on the basis of the territorial expanse of the empire before that fatal day in 1389, nor is historical Great Serbia what the Serbs currently aim at, either. Any attempt to recover this ancient empire would produce an international conflict of unmanageable proportions. Currently 'Great Serbia' means the joining of the dispersed Serbian settlements in Montenegro, Serbia itself, Bosnia and Croatia – a programme difficult enough in itself.

Figure 6 'A house built on co-operation' ('Catholicize!' 'No, circumsize!')
(Milenko Mihajlović, 1989) (from Banac, 1992)

Most historic events determining current Serbian reflexes refer to a much more recent past than the Great Serbian empire. But even for this we go back to the beginning of the nineteenth century, when a more or less autonomous Serbia re-emerged (1830). This period in history has yielded at least three themes that have entered profoundly into Serbian perceptions of the outside world: the final struggle against foreign (Turkish) rule, with its traditions of revenge and mutual treason; Serbia as a potential arsonist in the world (that is, Serbia as possessor of mysterious power and powerful protectors); and Serbia as one of the world's tragic nations that has suffered from genocide.

Figure 7 A Catholic prelate with the U-symbol of the Ustasha, telling his rosary of plucked-out eyes (Milenko Mihajlović, 1989) (from Banac, 1992)

Under Ottoman rule it was not unusual for resistance to be cruelly avenged. Atrocities such as that of 1876, when more than ten thousand Bulgarian villagers were butchered by bands of Turkish irregular soldiers (bashibazooks) were encouraged by the decline of the power centre in Istanbul. The historic context of the struggle against Muslim Ottoman rule produced the theme of a nation that gave its blood in defence of a Christian Europe. There are other nations in Europe which derive this kind of frontier feeling from events in their own histories (the Poles, Hungarians, Russians and Croatians for example). The absence of any recognition of this historic role, and the sense that one is even considered oneself to be part of the barbarian periphery of Europe, reinforce Serbian bitterness at international criticism.

A quite different question is whether the Ottoman past has led to structural effects in the national character and social organization that make the

Balkans an unstable area for a long time to come. There is a concept in organization theory that says that organizations usually adopt some of the features of the environment with which they are struggling.[9] This means that, besides the attitudes of civil defense (militias) that stem from the anti-Turkish resistance, a certain warlordism with its organization into clans, leadership aiming at personal enrichment and lack of a separate economic interest in the bureaucratic class, could have been passed down to the society of the newly independent Serbia in 1830. Some of the Serbian groups, both inside (in Vojvodina) and outside the territory of lesser Yugoslavia (Croatia, Bosnia), had known a different history because they had come to live within the sphere of influence of the Habsburg empire. These Krajna Serbs (Krajna = frontier) had in the nineteenth century no recent experience of Turkish rule but they were, although living as peasants, armed by the Habsburg government to guard the frontier. Thus they shared with other Serbs the interlocking of military attitudes and civilian life.

One may be inclined to recognize the continuity of such influences in the vicissitudes of the Serbian state since independence, in the competence for power, the undemocratic behaviour of leaders (King Milan), the scandalous events at the Serbian court around 1900[10] or the tolerance of unofficial instruments of power like the 'Black Hand' society. The Black Hand, mentioned in the quotations from Edith Durham at the beginning of this chapter, was led by a colonel (Dimitrijević) and aimed at serving the Great Serbian ideal by non-democratic means and by agitation among the Serbs outside Serbia.[11] However, as Gale Stokes remarks, ' . . . observers from larger countries tend to forget the similar spectacles in their own histories when they smile indulgently at the passions of the small and the weak.'[12] The British King William IV acted more like King Milan than as a constitutional monarch. And, would we, without the wisdom of hindsight, have forecast a democratic future for France with its history of revolutionary terror, or for Germany after the events of the twentieth century? Serbia should in fact surprise us by its autonomous adoption of a modern political system, with a constitution, parliament and functioning political parties, in the years after 1860. Representation of the people and the mitigation of absolute monarchical power was realized in Serbia before the country was modernized in other (economic, social) ways. The nation itself, however, remained a backward peasantry without a significant middle class or an aristocratic elite. This situation contributed to a politics that could find no other outlet than emotional nationalism praising the culture and dignity of the fatherland or radical attack on the state bureaucracy. As in southern Italy, as analysed by Gramsci,[13] or in Argentina (see Chapter 6), politics often dealt with politics – the (re)distribution of power, the constitution, the state bureaucracy – rather than with directing social and economic development itself. This situation might have evolved in a more harmonious way in the course of time if this part of Europe had not become entangled in so many wars. The severity and frequency of these wars was not entirely a Balkan feature: there was always an added dynamic from powers outside the Balkans.

The fate of Serbia has been determined for more than a century by the Great

Powers, who used the Balkans as a playground for balance of power politics. The most threatening of them were initially the Austro-Hungarian monarchy and, later, Germany. But Serbians also consider Italy, Bulgaria and Albania as states which will never miss an opportunity to play Serbia a nasty trick. The murder of the Austrian Archduke Franz Ferdinand and his wife in 1914 by a Serbian nationalist was the immediate cause of the First World War. Austria wanted to teach Serbia a lesson, but it unleashed a 'worldwide' conflict in which it perished itself. In the history of a nation such events cause such deep impressions that they excessively bias later interpretations of the world (this is what is sometimes called a 'peak-experience'). The first implicit lesson of 1914 was that the entire world would shake with fear at anything happening in and around Serbia. The second message was that anyone assaulting Serbia would be punished by an evil fate. If we can trust Edith Durham, this idea of being invincible already existed before the First World War. Some people exclaimed:

> We – the Serb people – have beaten the Turk. We are now a danger to Europe. We shall take what we please. The Serbs will go to Vienna. We shall go to Sarajevo. We have the whole of the Russian army with us. If you do not believe it – you will see. We shall begin in Bosnia!'[14]

With regard to the myth of invincibility, an illustrious history can mainly be assigned to the Montenegrins. Montenegro was never really conquered by the Turks and was consequently in a more or less permanent state of war. This induced them to look far outside the Balkans for support. Trade contacts with Western Europe developed early and this may explain the willingness of Montenegrins to give in to external pressure in the current conflict. The most realistic allusion to military strength is, of course, the partisan resistance in the Second World War. The armed units operating currently in Croatia and Bosnia call themselves Chetniks after the nationalistic partisan group of Mihajlović. By reviving such names they identify their opponents (particularly the Croats) as fascists and invest themselves with the reputation of being untameable and invincible. It is a message directed at both the traditional antagonists – Croats and Muslims – and the European powers.

As a theme that refers back to the experiences of the last world war, genocide seems currently more central to official Serbian policy than reminiscences of the partisan movement. According to 'scientific' publications of the Serbian government, the Serbs have the right to an ethnically pure territorial unit – something on which their 'defensive instinct' is also focused – because they have been subjected to ethnic extermination. That this is no romantic longing for purity is demonstrated, according to the document, by the current 'repetition of genocide of the Serbian nation in Croatia'.[15] There is, however, no historic evidence for a tradition of genocide of the Serbian nation.[16] The events of the Second World War are probably the most ambiguous in this respect because Serbians were indeed killed by Croatian fascists, but it is difficult to single out such events from the overall picture of violence between nations in this war. Nonetheless, official statements depict the Serbs as one of the 'tragic nations' that have been exposed to 'extraordinary suffering during wars and revolutions, like the Jews, Armenians, Kurds, Vietnamese, Polish and the majority of the Soviet peoples'.[17]

As the text of this chapter was originally written and published in 1993 and perspectives in the Yugoslav war have been changing ever since, I shall take advantage of this opportunity to update the previous conclusion, not because it has been painfully denounced by new facts. On the contrary, events leading up to the Dayton Accord have confirmed my basic assumption that any successful international action should take account of Serbian 'iconography' which I have specified as invincibility in war, the evil fate of any power meddling with Serbian matters (on impending NATO attacks general Mladić even invoked a parallel between Serbia and Vietnam), and Serbia as one of the world's tragic nations. As long as international action (unwittingly) affirms such visions one cannot hope for compliance by a warrior nation. We have seen, however, that direct military action, the absence of international repercussions (such as Russian help for Serbia), and international recognition of Serbian war crimes, have shattered Serbian myths, which has possibly been more decisive than the shattering of Serbian military *matériel*.

It is not the correctness of this conclusion that needs to be emphasized in 1996 but the limitations of its scope. I have noticed that it is impossible to deal with a limited aspect of a complicated human tragedy like the Yugoslav conflict without becoming involved, or assigned a position, in intellectual discussions. The first of the ordeals which were impossible to escape in 1993 was being branded as 'demonizing Serbia'. Focusing on one national identity at a time suits the approach of this book but in no way does it imply that one is putting the blame for a war on one side only. Even though I do not hesitate to point to the Serbians as the party that has committed the majority of atrocities in this war, I want to emphasize that the question of culpability is extremely difficult to answer in this case. In a chain reaction of fear, all acts of violence are understandable to a certain extent. However, writing in 1996, I am rather astonished that the theme of 'demonizing' Serbia has so completely disappeared from fashionable discussion.

A more empirical objection is that the Serbs are treated as one actor whereas one should at least make a distinction between those Serbs from 'Little Yugoslavia' and the scattered Serbs in Bosnia and Croatia. The apparent compliance of Little Yugoslavia with international pressure and the irreconcilability of the Serbs behind Mladić and Karadžić seemed to confirm this view. However, subsequent developments, such as the emergence of Serbian President Milosević as the unified Serbian voice in the negotiations, revealed the mutual understanding between Belgrade and the Bosnian and Croatian Serbs. This is not to deny that there may be great differences in interest and in existential feelings of insecurity between both groups.

Then there is the vexed question of how to understand 'irrational' violence itself. Geopolitical visions help to understand why the Serbs are aiming at ethnic purity and why they are so stubbornly pursuing their goals in the face of opposition from almost the entire outside world. But do these visions also animate the sniper who plugs a teenager, do they prescribe the rape of women? There is a very simple answer to such questions: raping, looting and aimless games with death have always accompanied war. An explanation of this requires the psychology of war rather than the political or cultural analysis of the ethnic groups of former Yugoslavia. However, many commentators adhere to the impression that such acts have a special significance in this case because they so clearly go beyond all rational military goals. The anthropologist Mattijs van de Port has suggested that this behaviour is a reaction to the collapse of all narratives creating the Yugoslavian sense of identity and dignity: the idea of being the esteemed leader of the non-aligned countries, of evolving into a normal European state, of a more comfortable life in the future, and so on.[19] Faced with failure at the 'periphery of civilization' after putting so much effort into keeping within those limits, the Serbs now want to break with all false narratives and to demonstrate to the rest of Europe and to each other: 'You thought you knew what reality was? Look at us. We have cast off all falsehood and this is what remains.'[20] Such an account might apply to other countries, such as Russia, which means either a depressing prospect for the nations concerned or – in the absence of similar degeneration elsewhere – the necessity to specify the special features of the Yugoslav case more extensively. One might speculate that Russia has sufficient 'weight' of its own to survive the failure of European integration. For the Yugoslav nationalities it means completely rewriting one's identity. If there is some truth in this analysis, then it is a great tragedy that a war unleashed because a certain 'narrative' (of Yugoslavia) failed, had to be stopped by destroying the (Serbian) myths that still remained.

TOTALLY LOST?

(Iraq)[1]

QADISIYYA

In his last attempts to avert a great battle, the Persian supreme commander Rustam asked the Arab troops camped beyond the Euphrates to send over a delegation of negotiators. The Arab commander, Sa'd ibn Abi Waqqas, did not want to do the enemy such an honour and sent an old Bedouin on a ragged mare. Rustam, seated on a throne, received the delegate with pomp and circumstance in a hall whose floor was covered with precious carpets. The old Arab, dressed in shabby clothes, was not intimidated and remained seated on his horse as long as possible. He walked the last few steps, leaning on a lance with which he carelessly pricked the precious carpets. When the guards made as if to seize him, he exclaimed, 'I have not come to this place to be disarmed but because I was invited. It was not my wish to appear. If you disarm me I will turn on my heels.' Rustam ordered his men to retreat and asked for the Arab's list of demands. His reply was that there were only three options for the Persians: conversion to Islam, becoming tributary to the Arabs, or war. After Rustam's refusal to comply, in the days that followed other delegates were sent with the same message. The supreme commander, realizing that further negotiation was useless, decided to open hostilities. It is the year 636,[2] and we are on the threshold of the Arab victory at the battle of Qadisiyya.

In September 1980 this place-name achieved prominence once more, when Saddam Hussein's Blitzkrieg against Iran seemed to slow down after a week. Whereas his initial declarations had referred to territorial claims – authority over the Shatt-al-Arab and a few offshore islands in the Persian Gulf – the war now was called a 'second Qadisiyya', or 'Saddam's Qadisiyya'.[3] A world of meanings and feelings was invoked by this simple reference: the identity of the victor, the power of simplicity and spontaneity against boasting and bureaucracy, the pan-Arabic character of the war, and others. International crises produce what some authors call 'moments of high drama which can illuminate the political landscape . . . like a flash of lightning.'[4] Was the war with Iran and the reference to Qadisiyya such an opportunity in which the geopolitical vision of the Iraqi people becomes suddenly illuminated? Is Qadisiyya a symbol of what binds the nation together (what is sometimes called a national 'iconography')? In such an unstable region as the Middle East the answer to such questions may be highly relevant.

In the first place, of course, we have to recognize that many tacit and explicit references, like those to the simplicity of the nomadic lifestyle, do not avow a particular Iraqi identity. These are suitable symbols to defame the enemy, either the West or Iran according to the situation, but they affirm an Arab rather than an Iraqi identity. They may even better suit a more general framework. Muslims consider wealth, show-business, sex and the absence of sober rules of living as signs of decay and weakness, which they prefer to associate with the West.

On the other hand, it is quite understandable that the many lateral connections between Islamic, Arab and local meanings make such myths useful for national purposes. Thus, shortly before the launch of Operation Desert Storm (in January 1991), Saddam Hussein mentioned his discovery that the elephant was the emblem of the American Republicans. His statement was received with great satisfaction by his military staff and considered a favourable omen for the coming battle.[5] Elephants can be associated with the battle of Qadisiyya in which the Persians used elephants, but there is also a connotation with the Ethiopian conquerors of South Arabia who in 570 set off for the Kaaba with an elephant, to destroy the sanctuary. As a result probably of an infectious disease, they had to retreat. In the Koran this event is commemorated in Sura 105 ('The Elephant'):

> Have you not heard how Allah dealt with the Army of the Elephant? Did He not foil their stratagem and send against them flocks of birds which pelted them with brickstone, so that they became like plants cropped by cattle?'

A variation of the Qadisiyya myth stresses the ability of Arab warriors to take the initiative, without being dependent on rules or an external authority (contrary to the Persians, who relied on 'bureaucratic' structures)[6]. However, this positive characteristic is withheld from some Arab brothers. The Iraqis consider the Egyptians to be a people with a slavish character.[7]

Islamic and pan-Arab references do not, at any rate, imply that there is no room left for a specific national 'iconography'. On the contrary, Islamic and pan-Arab symbols are even dug up by various regimes in the Middle East to preserve national unity, threatened by several forces but not by an emerging Muslim or Arab awareness.[8] In this chapter the focus is on how this process has evolved in Iraq and on the possible influence of a basic national iconography.

SADDAM AND NEBUCHADNEZZAR

During the eight years of the Iran–Iraq War (1980–1988) the word Qadisiyya emerges time and again in words and pictures. Large outdoor paintings, postage stamps and illustrations in newspapers and magazines show Saddam Hussein as commander in the classic battle of Qadisiyya, sometimes leading a column of charging tanks at the same time.[9] A postage stamp from the 1980s with the caption (in English) 'The First Qadisiyya Battle' also pictures Saddam Hussein, who is casting a shadow in the shape of Sa'd ibn Abi Waqqas.[10] But there are also other historic references. In the discourse of the President and his Ba'th Party, the pan-Arab theme has been slowly displaced by the national interest since 1970. The Iraqi leaders did not recoil from revaluing the pre-Islamic era, although this was less candidly expressed in political statements than in domestic cultural policy. Archaeological research, for example, was subsidized very generously. The director of the Babylon excavations told an American journalist in 1979' 'Whatever we want, we get . . . Not just a million or two, but anything we wish, without restrictions!'[11]

At a time of economic austerity during the war years, these excavations could also count on a 'blank cheque' in the name of the president because Babylon had to remain 'an inspiration for the people in its fight against Iran'. Another manifestation of this attempt to revive the ancient past can be found in the naming of official organizations and objects. Since 1970 the Iraqi regime has renamed several administrative districts. One province was renamed Qadisiyya, and two others got the names Nineveh and Babylon. Symbols of Babylonian, Akkadian, Assyrian and Sumerian origin are to be found on coins, postage stamps, decorations and emblems of organizations.

Strictly speaking, such expressions of territorial nationalism are at odds with those two other ideals which the Iraqi leader and the Ba'th Party endorse: pan-Arab nationalism and Islamic internationalism. The former considers the boundaries dividing the Arab world as a product of continuing imperialist manipulation, preventing the Arabs from reaching their true strength and finding their true selves. Islamic internationalism reduces the meaning of national units entirely to practical administrative functions. Or, as V.S. Naipaul, in search of the practical application of Islam in the state, remarks about one of his conversation partners, 'His faith was so great that he could separate his country from its history, its traditions and art: its particularity.'[12]

For both pan-Arab ideology and Islamic internationalism a particular 'national' culture and identity only become real with the acts and teachings of the Prophet and the subsequent Islamic expansion. Compared with Nebuchadnezzar, the symbol of Qadisiyya is completely untainted. The ideologists of the Ba'th Party never went so far as to hail the ancient Mesopotamian civilizations as 'Arabic'.[13] For Saddam Hussein these identifications were nonetheless useful. First, because even Muslims are susceptible to characteristics which distinguish them in a positive way from the world community. Second, because the all-embracing problem in Iraq is the social-cultural partition into three groups, Kurds, Shiites and Sunnis. The Mesopotamian heritage suggests a kind of common bond between these groups that cannot be found in other ideologies and does not cover other countries either. By linking up with a spontaneous movement among modern artists in Iraq (the use of old-style forms and symbols in a kind of modern primitivism) and by suppressing the pre-Islamic identity in propaganda to the outside world, critical reactions from other Arab countries were avoided. Sometimes the pan-Arab cause could be very well expressed in these symbols. During the Spring festival in Mosul in 1971, Nebuchadnezzar (dressed as an Iraqi) was shown at the head of an army leading along the Jews taken captive in Jerusalem, as a 'lesson to Zionism'.[14]

Figure 8 Saddam Hussein and Nebuchadnezzar (Mahmud Hamad, 1987)
(from Baram, 1991)

One may question whether Saddam Hussein has been successful in his attempts to cultivate a sense of national identity. The effect of all the historic symbolism was not that the Iraqi people became proud of their common past. Rather, it made the person of the leader even more threatening and intangible: most references to the Mesopotamian past show Saddam Hussein in a prominent role. He is the personification of all national norms; outside of him no rules for national policy-making exist. After a conflict with Iran that is still unfinished and that has not brought the victory that has been conjured up before the people, the great leader starts a new military offensive against a neighbour. To keep a free hand, he unilaterally offers to Iran to withdraw to the frontier of the (Algiers) treaty of 1975 – thus making all the sacrifices of eight years of war in vain – and announces the 'mother of all battles' against the superior army which has come to liberate Kuwait. Defeat in the new war is absolute. In ten years of war 400,000 Iraqi soldiers have possibly been killed in battle, and the direct and indirect national economic damage may amount to a thousand billion dollars. Not even a slightly democratic government would have survived such a failure. But Saddam lives.

REGIONAL RATIONALITY

Irrational as the actions of the leader may be in his attempts to enforce a national identity, it would be an error to judge his foreign policy as the expression of a pathological personality. The background of Saddam Hussein's actions comprised a geopolitical vision that could depend on the Arab world for recognition and had a significance wider than just within the small circle of the Iraqi leadership. Although the closed character of this circle and the distrust of even his closest relatives may seem to justify the term 'groupthink', the President's appreciation of the mood in many Arab countries was not other-worldly. Groupthink denotes the loss of critical sense and the excessive self-confidence in small groups of decision-makers who have cut off all inconvenient information from outside the group. However, in the regional (Arab) context the Iraqi policy was based on a relatively good assessment of political strengths and weaknesses. The decision-making that finally resulted in the Gulf War relied on a geopolitical theory of which the outlines (in *italic*) were as follows:

(a) *The Arab countries, Iraq in particular, are threatened from one side by Israel and from the other by Iran. Imperialist forces are bent on organizing an Iranian emigration wave, as they have done in Palestine.*[15] This observation possibly attests to the pathological fantasy of conspiracies (of which the Arab intellectual world provides many examples),[16] but the image of the aggressive posture of Iran is not unfounded. The Shah was open about his aim of making Iran the regional military superpower that would protect the oil interests of the West. It is also true that there were trade relations between Israel and Iran for strategic equipment.

(b) *Iraq has protected the Arab world against an Iranian advance at the expense of countless human lives and immeasurable financial sacrifices. This qualifies the country as a natural leader in the Arab fight on the other front (against Israel), whereas the other Arab states have forfeited their right to leadership. This morally entitles Iraq to financial support from these states.*[17] (See Figure 9.)

(c) *Considering the above, the unwillingness of Kuwait to cancel Iraqi debts, the overstepping of its allocated share in the production of the Rumaila oilfield, and the fall in oil prices through overproduction during the war with Iran, the demeanour of Kuwait is evidence of a lack of Arab solidarity.*

Figure 9 Saddam Hussein as benefactor to the Arab world (Mahmud Hamad, 1987)
(from Baram, 1991)

If the Iraqi leadership, and particularly the president, were thinking that they had a strong case with these arguments, they were right. The only uncertainty was whether these arguments could neutralize the negative reactions that an invasion in Kuwait would certainly elicit. The generally accepted rhetoric that the boundaries between the Arab countries are only a product of foreign interference and consequently invalid, did not justify a unilateral intervention, let alone violence against brothers. Anything of that kind required at least broad Arab consultation. But the conclusion that a basis for support existed, particularly among the Arab masses who disputed the unequal distribution of wealth from oil production, was also right. Annexation of Kuwait was possibly 'not legal but certainly legitimate' as several Arab intellectuals commented. It was to be expected that massive support would keep the autocratic rulers of countries like Saudi Arabia from retaliating. There would be talks in the Arab way, endlessly, and finally the situation would become accepted. In this respect Saddam's appraisal was not bad at all.

But what of the Americans? Saddam Hussein's view of the risk he ran from that quarter was clearly expressed in the conversation he had with US ambassador April Glaspie on the eve of the invasion (25 July 1990): 'Yours is a society which cannot accept 10,000 dead in one battle.' – in spite of all its cynicism (namely that Iraq could accept so many deaths), a correct assessment. Saddam's only error of judgement was that he did not realize how far the new conditions of war made it possible to reduce the number of casualties (at least on the American side) by the deployment of high-technology equipment. And, of course, that the United States, in spite of the high costs, wanted to set a powerful example to prevent further instability in the region.

But even if it did come down to a foreign intervention, the scenario was not hopeless. Western interference would immediately cause Arab ranks to close and accentuate the front-line position of Iraq. That battle would be won even if it did prove necessary to withdraw from Kuwait. From public opinion research in Arab countries and from reports of correspondents it can be inferred that these assumptions were not far off the mark. Immediately after the invasion of Kuwait a majority of Arabs was inclined to denounce Saddam as instigator of a counterproductive international crisis. But after the Gulf War a majority thought that events in and around Kuwait had been just a pretext to destroy Iraq and humiliate the Arab world.[18] Or, in a more radical variation: The invasion of Kuwait was even instigated by the United States.

The regional level was the only one in the Arab world at which a real political culture could manifest itself, as a unity of political goals – a 'discourse' – connected with generally accepted values and enjoying a large measure of political legitimacy. The same cannot be said for the *domestic* politics of the states concerned. The Gulf War caused such a drastic rupture in the unity of political discourse and practical policy that the regionally derived legitimacy of the separate states has been lost. The Ismaels suggest that the foreign relations of these states are beginning to resemble their domestic politics.[19] By this they mean that a kind of pure power politics emerges that is no longer alleviated by a common culture and norms. The consequences are difficult to foretell. Chaos theory has it that turbulent situations have the potential to start developments in opposing directions from the same initial state of affairs. In any case, from now on these states are thrown back upon themselves, which redirects the question of stability to the problem of domestic rationality.

DOMESTIC RATIONALITY

Stable states are based on a 'state-idea', a generally accepted argument answering the question why certain boundaries enclose a particular community. This is approximately what the political geographer Richard Hartshorne stated in 1950.[20] But where is the state-idea formulated? To illustrate the difficulties involved, Hartshorne mentions a case which had put one of his graduate students to great trouble: Iraq. After much reflection, the student managed to produce the formula: The state-idea of Iraq is, 1) the Great Powers' acknowledgment of the special strategic and economical significance of Mesopotamia; and, 2) the need to provide Arab nationalism, exiled from Syria, with a *pied-à-terre*. These starting-points led to the decision to delineate a territory that connected the densely populated Arab region in the Euphrates and Tigris region with the bordering, but dissimilar regions populated by mountain and desert tribes. This solution, Hartshorne sceptically remarks, is a mixture of primarily external objectives from 1919. Now (he was writing in 1950) one would have to check whether Iraq had become a reality in the hearts and minds of the inhabitants. Almost half a century later this question is still there and all the more relevant. The answer has been given without reservations: 'The mosaic of communities and sects that make up the country have never been prepared to sink their differences into a broader sense of community, as happened, for example, in Egypt.'[21] According to Samir al-Khalil (Kanan Makiya), Iraqi nationalism in the sense of identification with a territory is out of the question. The spark of community spirit that still existed during the war with Iran has effectively been destroyed by Saddam Hussein's terror campaigns against the Kurds and Shiites after both wars (in 1988 after the war with Iran and in 1991 after the Gulf War).

Khalil explains that Iraq has developed into a pathological regime of fear that the citizen confronts even in his personal life, with unexpected re-organizations, liquidations, and other events that are never officially cleared up. Only ambiguous messages emerge about imperialist conspiracies and subversive forces. In such a system it is impossible to test public events rationally against stable norms and expectations. Everything can suddenly change its meaning[22] and the paradoxical result is that unfulfilled official promises and theories of the world that are contradicted do not injure the authority of the leader. The vulnerability that the citizen senses in this system expresses itself in a peculiar pathology: 'Doomed to teeter on the edge of the precipice, they are possessed by the need of a safety line of some sort. Hero worship of the Great Leader variety presents itself as such a safeguard.'[23]

These observations require comment: one needs to distinguish between the particular effects of Khalil's 'republic of fear' and the generally accepted 'statist' approaches to politics in the Middle East. 'Public opinion' in Arab countries allows governments to interfere deeply with society, even while limiting human rights and freedom. State control is seen as the only effective means to escape the condition of underdevelopment and the embrace of imperialism, a task which is conceived almost in military terms and images. In Arab countries the public sector consequently has a very big share in the national economy.[24] Apart from the 'battle against underdevelopment', Islam is identified as another cause of the absence of democratic institutions. A society that rejects all worldly (say Western) institutions because no positive legitimation can be found in religion, creates a void that will be easily filled by the most well-organized and most alert social institution that remains, Naipaul opines.[25]

Current developments suggest that if the public loses confidence within the Arab world, it concerns confidence in the power of governments rather than in the power of Islam. In the Arab world one also begins to acknowledge that the 'public sector' is not the most effective means to safeguard national wealth. This loss of confidence proceeds slowly, because earnings in oil production blur the inefficacy of economic systems. But in countries without such revenues, young people embittered over lack of opportunities may easily turn to fundamentalism.

Regarding revenues, the prospects for a future regime in Iraq are probably better than for one in Egypt or Turkey, but one may wonder whether an Iraqi state (on a less criminal base) will get the opportunity to grow in the minds and hearts of the citizens after all that has happened in the last decade. On the one hand, a retreating state implies a severe threat to political control. On the other, it is possible that a state that interferes less in the daily life of its citizens may be accepted as a practical necessity in the long run. People can live without strong identification with a state if they are themselves respected and left in peace. This requires, first, a reinforcement of economic ties between small entrepreneurs and the diverse parts of the national territory. The state should not play a dominant part in such a process. Perhaps the state will have to remain strong, even autocratic, in its international relations, to guarantee national security. But the identity of the Iraqi citizen will have to be rooted in the culture of daily life. It is to be hoped that grand geopolitical visions will not intervene in that process.

A WORLD IN ITSELF

(*India*)

INDIA'S UNITY AND DIVERSITY

Geopolitics has now become the anchor of the realist and its jargon of 'heartland' and 'rimland' is supposed to throw light on the mystery of national growth and decay. Originating in England (or was it Scotland?), it became the guiding light of the nazis, fed their dreams and ambitions of world domination, and led them to disaster . . . And now even the United States of America are told by Professor Spykman, in his last testament, that they are in danger of encirclement, that they should ally themselves with a 'rimland' nation, that in any event they should not prevent the 'heartland' (which means now the USSR) from uniting with the rimland.[1]

In these terms India's first Prime Minister, Jawaharlal Nehru (1889–1964), writing in prison in 1945, expressed his concern about the possible recurrence of imperialist strategies in the postwar era. From his subordinate geographical position in the colonial 'rimland', Nehru anticipated the intellectual forces and fears, soon to be unleashed as the Second World War neared its end, better than many a statesman in one of the victorious countries. He seems less concerned about the easy transfer of Mackinder's thoughts from England to its enemy Germany – which is all old Europe, after all – than he does about the emergence of kindred ideas in the United States. At the same time Nehru proclaimed his belief that a geopolitical perspective in international relations is actually outworn. His reflections rely on a curious mixture of realistic and more philosophical assumptions. The realistic cause for concern is Nehru's acute awareness that in the near future India will have to deal with two new great powers, the United States and the Soviet Union, which are still at odds in their world policy outlook. The philosophical heart of Nehru's ponderings is the assumption that the postwar world will display a qualitative change rather than a revival of the old balance of power game with just a change of players. The new world order, as Nehru saw it, would no longer be dictated simply by assemblies of Great Power representatives whose words are pressed home by force of arms. Things could only be changed if the 'feelings and urges of vast numbers of people' were taken into account. Self-interest, if based exclusively on national gain, was not rewarding any more. In order to maintain vitality and wealth, states would have to co-operate in future.

Nehru was not writing about India's coming international relations in particular, for his thoughts primarily concerned the domestic situation. Convinced of the unity of India (within its British imperial borders), Nehru had to face Muslim pressure for a separate state. Such ideas about a future division of India are treated to the same criticism that Nehru levels at traditional 'realist' power politics: the small nation-state is a phenomenon of the past and a territorial division of India would soon reveal how dependent both new units are on each other and would immediately raise the need for a federal association. Besides, any acceptable territorial division would leave the Muslims with a territory that was both smaller and economically less viable. The prospect for the Indian subcontinent was either 'union plus independence or disunion plus dependence'.[2]

Nehru's conception of India rests on its supposed unity of culture and on what Ashutosh Varshney has called a 'sacred geography' in a metaphorical, not in a literal, sense.[3] A literal conception of sacred geography would convert India's territory into a holy land, a country strewn with religious monuments and places full of echoes of saints and sacred events. This is how Hindu nationalism conceives of India's unity, but Nehru's conception is a secular attachment to its diversity and pattern, to a 'geographical entity' that he never describes in terms of natural laws or boundaries.[4] The implication is that there must be something else, an historical experience, or a common culture, delineating the community that appropriates its common geography as something beloved. As Nehru states in The Discovery of India,

Some kind of a dream of unity has occupied the mind of India since the dawn of civilization. That unity was not conceived as something imposed from outside, a standardization of externals or even of beliefs. It was something deeper and, within its fold, the widest tolerance of belief and custom was practised and every variety acknowledged and even encouraged.[5]

But if such a deep feeling existed, one has to acknowledge frankly that it was not shared by everyone. Two years after Nehru wrote these lines, a rage went through India, leaving a million people dead and ten million refugees who suddenly found themselves crossing borders in what had previously been a single country. The tragic events of the separation of India and Pakistan suggest that Nehru was actually engaged in 'constructing' rather than 'discovering' India. He might even have succeeded in transferring his concept of India to the masses if the newly independent people had been given some time to reflect. After all, India still exists, somewhat reduced in its size but still with enough internal religious antagonisms to raise frequently the most gloomy forecasts, particularly in Western commentaries.

What Nehru sought to overcome in his historical perspective spanning 2,500 years or more, was 200 years of British rule. His idea of unity was nonetheless conditioned by the era he wanted to wipe out. The colonial experience made itself felt in India's geographical delineation and in the irresistible explanations or justifications it provided for religious antagonisms within India. Before independence, nationalists usually blamed the frequent occurrence of Hindu–Muslim riots on the imperialist system. The explanation was either that the riots were deliberately provoked in order to prove that the Indians were incapable of taking care of the country, or that the prevailing system of economic exploitation caused extreme feelings of relative deprivation between members of different groups. Such angers and passions would automatically disappear with the coming of political independence, Nehru thought. People would come to understand that differences in social position and wealth were cutting through religious groups and were a common challenge rather than a sign of religious discrimination. It was not to be. As Embree says, Nehru was unaware 'that the same democratic processes involved in building the new social and political order for which he laboured could work to strengthen the divisive religious forces that he very genuinely deplored.'[6]

Obviously national identity must have different meanings for different groups in India. The large majority (85 per cent) of India's 900 million inhabitants, the Hindus, have the widest range of options. The individualist nature of their religion admits secular conceptions of the state like those advanced by Nehru and his Congress Party, in both a territorial and a cultural sense, but they may conceive of their state as a holy land as well. Today's new Hindu nationalism, the Hindu radical Right, is stressing the indispensability of precisely that latter viewpoint. It argues that territory, state apparatus and constitution do not in themselves legitimate the Indian state. Legitimation and identity can only be derived from the source of India's culture, that is, Hinduism. Antagonism with the largest minority, the Muslims (about 12 per cent of the population) is a logical extension of this position. The destruction in 1992 of the Babri Mosque in Ayodhya, claimed by Hindus as a former temple devoted to Ram, was one violent outcome of the new nationalism in recent years.

The Muslims are in a more complicated position with regard to India's national identity. Their contribution to India's culture, architecture and literature is conspicuous but they cannot claim Indian territory as a holy land. Neither is their religion as easily separable from community matters and institutions, such as marriage, personal law and property rights, as in the Hindu case. This has caused profound conflict with the secular principles of the Indian state. Finally, the simultaneous creation of a Muslim state and the independence of India meant that Muslims in India might either become susceptible to conflicts of loyalty or run the risk of being regarded as a fifth column. One response has been to create a movement similar to the Hindu radical Right, the Jama'at-i-Islami, which aims at creating an Islamic state in India. The idea is devoid of any realism, of course, but realism is not the stuff religious (and desperate) politics is made of.

Other movements bearing on India's national unity are the separatist nationalist movements in Punjab (Sikhs) and Kashmir (Muslims) and the regionalist Dravidian movement in Tamil Nadu. Although the separatist movements in the north concern small minorities, their impact on national politics is huge, first, because they threaten India's 'sacred geography' and, second, because foreign actors may be involved (like Pakistan in Kashmir and

Punjab), introducing a dangerous element of external insecurity. The Dravidian movement also obtained an international dimension with its connection with Tamil separatism in Sri Lanka (the Tamil Tigers). The dark side of these Indian 'fringe' problems was dramatically emphasized by the murder of two prime ministers who had dared to interfere with events in Punjab and Sri Lanka. Indira Gandhi ordered a massive attack on the Golden Temple in Amritsar in 1984, after it had become the refuge of a Sikh fundamentalist who commanded an armed band of supporters from this place. Some months later, Mrs Gandhi was killed by her own Sikh bodyguards. Her successor and son Rajiv was killed in 1991 by Tamil Tigers after the Indian army had assisted the Sri Lanka government in suppressing the Tamil insurgency.

Internal antagonisms, the sheer size of India's population and territory, and the daily burden of material survival, shape a subcontinental world whose 'internal' problems absorb most of the political energy and interest that Indians can produce. The idea of an Indian community with a definite vision of itself and a mission in the world is an illusion that can only emerge in places outside the subcontinent, which inevitably entertain a selective relationship with India. V.S. Naipaul tells how, on his first visit to India, his expectations were actually shaped by his reminiscences of Trinidad's Indian community. But the idea of a continent-wide identity, based on the independence movement and on the India of the great names and the great civilizations, only made sense when the community was very small, a minority, and isolated.

> In the torrent of India, with its hundreds of millions, where the threat was of chaos and the void, that continental idea was no comfort at all. People needed to hold on to smaller ideas of who and what they were; they found stability in the smaller groupings of region, clan, caste, family.[7]

Foreign countries and international relations are almost completely absent in the countless conversations Naipaul held with people all over India. The educational visit in London of a young Indian, and listening to Peking radio in the days of the Marxist (Naxalite) actions in the 1970s, are notable exceptions. During Naipaul's stay in India the Chinese students held their dramatic demonstration on Tiananmen Square in Peking (June 1989). There is no trace in his book of any comment or desire to discuss this issue which might have appealed so much to the Indian sense of democracy and appreciation of 'people's power'. Yet, there has been a special Indian note in the world of international relations and certain events in the world have reverberated in rather wide circles in India.

INDIA'S FALL FROM GLOBAL GRACE

India started its independent existence with a foreign policy that was global in reach and ambition. This comparatively prominent international role of the new state can be attributed to three factors: the personality of its first leader,

Nehru; the international experience to which the Indian bureaucracy had been exposed under British administration; and the geographical vastness and geopolitical situation of India.[8]

As *The Discovery of India* and other writings witness, Nehru was a man who had profoundly reflected on the world and who was able to draw original conclusions on global changes and events. India's international role until the beginning of the 1960s was outlined by Nehru alone, and nobody belonging to the domestic policy elite really challenged it. The pillars of Nehru's foreign policy were anti-colonialism and non-alignment.[9] These principles were almost generally accepted, although parliament was usually very much interested in foreign policy. As long as it brought India international prestige, which it usually did in the 1950s, everyone was satisfied. Nehru was unique among Third World leaders in putting the issue of Indian independence in a world perspective by 'reiterating that the freedom of India is a part of the freedom of the world . . . and that the problem of discrimination, racial or otherwise . . . constitutes but a part of the problem of racial and other discriminations prevalent in the world as a whole.'[10] Nehru believed that India could not play a secondary part in the world: 'She will either count for a great deal or not count at all.'[11]

The reservations of Indian leaders about (particularly) the Western world were not only induced by the colonial past. The vain destruction of lives and property in the battles of the First World War had convinced Mahatma Gandhi (1869–1948) early in the century that there was something 'satanic' in Western civilization. Gandhi concluded that the entire technical achievement of the West only ended in self-destruction. This view inspired the search for national strength in a simple, self-sufficient way of life. However, on this question the opinion of the two great Indian leaders diverged. Nehru basically accepted the principle of non-violence in world relations but his appreciation of the 'modern project', of Western science and technology was precisely the reverse of Gandhi's. Science was both a precondition for Indian independence and something which suited the Indian mind very well.

India's authentic and neutral position in world affairs and the prestige it acquired from mediating in international conflicts (for example, in Korea in 1949 and Indochina in 1954) were not replicated in its relations with neighbouring countries. The bitter separation with Pakistan should have been a warning that even an inclusive and peaceful world view will eventually meet geopolitical limitations. In October 1947, two months after independence, Pakistani troops ('volunteers') raided the territory of Kashmir, until their progress was checked by a hastily assembled Indian army. India concentrated troops in the area but did not switch to a more military approach in its overall foreign politics. It resorted rather to UN mediation, which succeeded in arranging a ceasefire agreement in 1949.

Relations with China provided another traumatic experience. They should have become a model for all relations between Third World 'brothers'. And indeed the two countries had much in common: they were the largest countries of the world in terms of population; they had recently fought the same struggle for independence; and they were both rooted in ancient civilizations. China was one of the few countries in which Nehru recognized the same 'vitality' as he did in the Indian people.[12] When China annexed Tibet in 1959 after nine years of occupation, Indian protests were weak and India's only support for Tibet was the offer to receive refugees. There was one common relict from the colonial past, however, which turned the brothers into enemies: India's northern frontier, the McMahon line, established by the British administration in 1914. After fighting between Indian and Chinese border guards broke out, China invaded India in 1962 and easily defeated the Indian troops. The Chinese could have occupied large Indian territories but they decided unilaterally to withdraw, meanwhile 'correcting' the borderline in the north-east and north-west.

China's act of aggression and India's helplessness was the most traumatic experience since independence. It completely shifted India's foreign policy priorities from being a global political presence to military defence against regional security threats. It also compelled India, faced with increasingly close ties between China and Pakistan, to enlist the aid of the Soviet Union, an act that evoked serious doubts about India's non-aligned status. These changes essentially meant the fiasco of Nehru's postwar line of foreign policy, as he himself admitted. During the last years of his life he actually gave the go-ahead for a dramatic policy shift. 'We must change our procedures from slow-moving methods of peacetime to those that produce results quickly. We must build up our military strength by all means at our disposal . . . Freedom can never be taken for granted.'[13]

While India resorted to Soviet help, the United States supported Pakistan because the country was identified as a 'domino' piece in the strategy of containment of the Soviet Union. In this way the South Asian region became subject to and divided by the same world power politics that a resistant but nevertheless participant India had always detested so much. The adversity in its international relations also affected India's status in the world. The message of the wars it could neither prevent nor win seemed to be that India was unable to solve its own problems and was in no position to act as an impartial arbiter in conflicts elsewhere in the world. India's international ambitions were scaled down to its immediate environment, where its self-confidence was somewhat restored by two subsequent wars with Pakistan. In the first (1965), Pakistan aimed at cutting off Kashmir by a surprise attack (called 'Operation Grand Slam' by its initiator Field Marshal Ayub Khan). The superior Indian forces, however, won a decisive victory and the army could even have marched on into Pakistani territory if external pressure had not forced the combatants to cease their war efforts. The myth of the invincibility of Muslim armies – inspired by British appreciation of Muslim and Sikh soldiers – was effectively destroyed. The next war with Pakistan (1971) took place mainly in East Pakistan (the later Bangladesh). It was provoked by Bengali discontent with their second-class political status in the country. India's intervention, after civil war broke out, was successful. The Pakistani army had to withdraw and a new independent state was born. India did not yield to Chinese and American threats, but the result was a deterioration in its relationship with the United States.

One could say that, faced with the geopolitical forces encroaching on the South Asian subcontinent, India had no other choice than to side with the Soviet Union. Although this never implied a surrender to the Soviet bloc, relations with the communist world were facilitated by a certain degree of ideological affinity. India's anti-Western, anti-imperialist mission since the struggle for independence, and the socialist or statist visions of national development of most Indian leaders until Rajiv Gandhi, did not quite constitute the groundwork for a neutral position. As an anti-colonial power the United States should have been omitted from India's definition of the imperialist world, but as the above quotations from Nehru's writings suggest, already at the start there was much Indian ambivalence about US geopolitical aims. And as long as India took a favourable, or at least neutral, position towards the United States, it was reciprocated with nervy anti-communist reflexes. The United States had always embraced the anti-colonial idea of self-determination, which implied a clear message of solidarity with Indian destiny. But after so many new states promptly turned to Marxism or socialism with a Marxist bent, Americans started to wonder 'what kind of gratitude that was', as Moynihan succinctly phrased the American question.[14] Just as they mistook the socialist declarations of many new regimes for a surrender to the communist bloc, the Americans failed to understand that from an ethnic perspective the Indian choice was curiously adaptive. Moynihan suggests that there is a clear analogy between the ideas of cosmic order, hierarchy and a rational universe which gave the Brahmins their sense of leadership and the Marxist ideas of social control and planning from above. Only with socialism-Marxism could the Brahmin caste create a niche for itself in a secular society. The challenge to both Brahmin dominance and socialist ideology came from the Janata Party (Bharatiya Janata Party), which rose to power in 1977 and represents particularly the Hindu middle class. The Janata Party criticized the foreign policy of preceding governments as something that hardly qualified as 'non-aligned'.

Another point of contention in relations between India and the United States was the question of nuclear weapons. India probably started to produce atomic weapons in 1985 and has refused to sign the Non-proliferation Treaty since.[15] As a consequence, India has been subject to American restrictions on provision of high-technology equipment (computers) and parts for ballistic missiles. US pressure even induced the Russian President Boris Yeltsin to cancel an agreed sale of missile engines in 1993. Since the disappearance of the Soviet Union and the Soviet retreat from Afghanistan, the strategic importance of Pakistan for the United States has diminished. This offers new opportunities for an improved relationship between India and the United States, which indeed currently seems to be developing in the field of economic co-operation.

Of all the hostilities between India and its neighbours, only those with Pakistan (particularly on the question of Kashmir) have remained close to ignition-point. Indira Gandhi made a great effort in the 1980s to cool down tensions with China and to re-open normal diplomatic relations. Her successor Rajiv continued this line of policy, which finally led to a provisional agreement in 1993 under Prime Minister Rao.

THE PUBLIC VISION

The political philosophy of statesmen such as Jawaharlal Nehru may bear the unmistakable stamp of a place and its history, but this does not mean that their thoughts provide a key to the mind of the general public, even its most educated members. A flexible mental picture of India's international situation would at least require a clear identification by the individual with the national state, but, as Naipaul discovered, such an idea is not highly relevant 'in the torrent of India'. Stanley Wolpert, extending his observation to the question of national security, draws much the same conclusion:

> To this day, in fact, bonds of 'National' unity are much less powerful for the overwhelming majority of Indians than those of familial, caste, local and provincial, or linguistic-regional ties . . . What it may portend for India's national security is another matter. There has, at any rate, never been any simple 'Indian mind' or 'Indian response' to external threats or challenges or, for that matter, to internal policies threatening change.[16]

Harish Kapur mentions some other impediments to the effective functioning of public opinion in India. First, Hindu cultural tradition would not be favourable to civic concern and participation: 'there is an absence of real interest in secular affairs beyond one's own concern.'[17] Second, the absence of any formal education among most of the Indian population makes it impossible to benefit from the national and international press. Third, international affairs are too complex to follow effectively, even for educated public opinion. The third obstacle does not exclude the possibility that educated circles have a crude image of how the 'outside' world is or ought to be organized. A few incisive facts will penetrate the smoke-screen of day-to-day worries and lack of information precisely because such facts must fit in, or clash, with a mental scheme, however primitive. That even crude ideas about international relations may only be found in a small circle of people does not make them irrelevant.

There have been indeed a few occasions when Indian (elitist) public opinion rose against the official line of foreign policy. Such cases provide an obvious indication of the existence of certain 'innate' collective visions of the world. It is significant that these occasions here almost always concerned Soviet incursions upon the freedom of other peoples. The first such event was the Soviet repression of the Hungarian uprising in 1956. When India abstained from voting on a UN resolution condemning Moscow's military intervention, 'the country' was outraged. Nehru was compelled to accept the argument that 'the Hungarian government was not a free government . . .'[18] The same happened to Indira Gandhi when Warsaw Pact countries invaded Czechoslovakia in 1968. However, the most severe public protest occurred after the 1979 Soviet invasion of Afghanistan. The government was compelled to correct the initial statements of India's representative at the UN.

An absence of foreign policy issues has also been found in the political discussions and party manifestos accompanying elections to the Lok Sabha (lower chamber of Parliament). Even in the lead-up to the elections of 1991, when the region of South Asia had just experienced a major war (the Gulf

War), and the break-up of the Cold War world order had become undeniable, no significant debate on India's role in the new world order was begun.[19] The Gulf War itself had certainly aroused interest and debate. The general picture of Iraq was favourable: a progressive Third World country trying to become independent of external financiers and governments, just like India. During the crisis Foreign Minister Gurjal (of the V.P. Singh government) travelled to Baghdad to have a cordial meeting with 'our friend' Saddam Hussein. The idea of Saddam Hussein's 'mother of all wars' was received in the press with much optimism about its chances of success. Nevertheless, the succeeding Chandra Shekhar government had conceded a refuelling facility for US warplanes on their way to and from the Gulf. Parliamentary opposition, particularly by its leader, Rajiv Gandhi, forced the government to repeal this accord. Maybe it was the confusing outcome of the war and the inability of both government and opposition to play a dignified role, that meant that in 1991 parties and politicians were still unsure how to conceptualize India's future international position.

Certain domestic political issues are difficult to separate from external relations (so-called 'intermestic' issues). For example, the demand for social justice may make it more difficult to comply with the requirements for obtaining IMF loans. The militant championing of the Hindu cause may negatively affect relations with Muslim countries. In the election campaigns of 1991, such debates, if they emerged, never alluded to international repercussions. The Bharatiya Janata Party's emphasis on Hinduization (*Hindutva*) as a means to restore national consensus was never seen as something that could at the same time 'enable Pakistan to justify its moral and material support to the separatists in those provinces who had been active in highlighting the sense of alienation of local Muslims and Sikhs.'[20] Partha Gosh concludes his analysis of party manifestos in the elections of 1991 and in the elections to the state assemblies in 1993 with some remarks on what he calls 'India's foreign policy apathy'. One of his explanations (and one that has not yet been mentioned) is the traditional development strategy chosen in India, with its emphasis on import substitution. The absence of a strong urge to integrate the Indian economy into the global system, and the identification of its interests with those of the Soviet bloc, led to 'a highly rhetorical stance against the West . . . and in support of Third World causes'.[21] This anti-Western and anti-American rhetoric may have stultified the public and intellectual interest in foreign politics. However, it never prevented successive Indian governments from taking very pragmatic decisions!

Another glance at this Indian mechanism of information processing is provided by press responses to the 'Chinese spring' in 1989. The television images and newspaper pictures of the events on Tiananmen Square made a lasting impression.[22] The Indian government, however, painstakingly adhered to the principle of non-interference in the internal affairs of another country, an attitude that was gratefully acknowledged by the Chinese government. The press widely reported the events, but neither the press, nor the government, seized the opportunity to stress the difference between India and China with regard to democracy and political culture. The policy of improving relations with China had apparently been accorded a higher priority than the upgrading of India's image in the West. One of the reasons for this attitude was the rapprochement between the Soviet Union and China. Gorbachev paid an official visit to Peking at the same time the Chinese troubles occurred. In view of Pakistan's alliance with China, India did not want to convey an impression of being politically isolated. Thus, India's reserve in commenting upon the Chinese policy may be explained as a clear sign of India's capability to pursue a pragmatic policy (*Realpolitik*), although it also fitted in with anti-Western sentiments. But what about the press and the opposition parties which on earlier occasions had so strongly resisted the government's indulgence of communist suppressions of freedom?

Jean-Luc Racine has suggested that it was difficult for the opposition (the BJP) to denounce the Chinese treatment of the dissidents because that party had just started a campaign against the democratic degeneration at home, that is, in the Congress Party and its leader Rajiv Gandhi.[23] With regard to domestic problems, India could not serve both as a model and as an object of criticism. The press was not subject to such dilemmas and China was not spared its criticism. 'A war against the people' and 'The great leap backward' ran some newspaper headlines. But in spite of their reliance on the big international press agencies, judgements in newspaper articles did not entirely reproduce the Western vision. The defenders of the Chinese cause (their messages mostly derived from the Xinhua press agency) were given more space than was customary in the Western press. Western disapproval of the Chinese policy even became an issue as such. It was criticized as hypocritical because Western powers like the United States and Britain had resisted international boycotts of South Africa. The Chinese, at least, were killing their own people, whereas Western countries preferred to kill other races. Only a few commentators pointed out that to brand the human rights movement and its criticism of China as a new Western means for suppressing the Third World, was a flawed argument. Human rights criticism could just as well be applied to the United States or Europe. But the general trend was to use the Chinese question to discuss the West.

These facts about the 'public' response to events abroad suggest that there is a tenacious anti-Western element in the Indian vision and a support of nationalist demands for freedom elsewhere in the world, but also a fear of dealing with the internal disturbances in other countries resulting from contested ideas of freedom. The precarious equilibrium of religious and other 'quasi-ethnic' groups (like the castes) in India may be one of the most effective impediments to the formation of policy visions on the domestic affairs of other countries. This is particularly true of the elite. Ashis Nandy has established a tendency in the Indian policy elite, which for that matter can also be found in other modern states, to distrust the collective spirit, a fear of populism and the 'emotional vulnerability of the ordinary citizen in international relations'.[24] In opposition to this aloofness of the elite, expectations of democracy are on the rise in the Indian masses, because they assume that democracy may provide them the freedom to pursue their own concepts of identity, even though these demands may ultimately endanger the stability of the state. The problem with

India is that these two developments, of the growth of institutional arrangements that the state allows to function by protecting its citizens, and the
increasingly explicit demands for freedom of the masses, have not yet come
together. As Nandy states: 'the Indian experience is uncommon in that this
relationship between democracy and freedom is as yet open.'[25]

A GEOPOLITICAL EPILOGUE

Since the fall of communism, or at least the demise of its leading state, and
since international political power has given way to transnational economic
mechanisms, the sparse geopolitical codes to which the wider Indian (educated)
public referred have become even more out of date. This suggests that 'India's
foreign policy apathy' may continue in the near future. However, this does
not necessarily imply the absence of pragmatic foreign policy decisions in
government circles. The concept of geopolitics has been introduced precisely
as the intellectual backbone of such pragmatic decision-making (although a
century of geopolitics has made us somewhat sceptical about this pretension).

Non-alignment has been the most distinct geopolitical stance that India has
adopted since Nehru's time. Whether it was really implemented, in its pure
form, is another matter. In the current world order the idea has lost much of
its significance.[26] Nevertheless, one might still see India's role in the world as
emanating 'from the country's unique geopolitical situation as a bridge
between the East and the West', as one author recently stated.[27] There are
many countries who claim this position, but in India's case it may have a
significance that transcends the meaning of being situated on the major transport axes between Europe and the Far East. India is the only important secular
democracy in the East and in this way combines features of the West and the
East (or the Third World). It is also an important regional power. In the Indian
Ocean region, India occupies the central position and is the largest littoral
state. These are undoubtedly valid arguments in support of its claim to a
permanent seat in the UN Security Council. The cultivation of regional power
status, however, can also be dangerous. It might draw two great neighbours,
India and China, into the abyss of what could easily become the most terrible
war of the coming century.

These realities still appeal very little to the Indian masses. Their feelings of
security are not primarily related to the world outside India. Even the Hindu
nationalists, the group which may most easily identify with the Indian state,
do not have a clear conception of the role of India in the world. The most
profound geopolitical reflex is still derived from India's colonial history and the
struggle for independence. In the absence of a new external threat, the only
national identity can come from the slow growth of a middle-class culture
with its film-stars, advertising, and other secular symbols of consumer
culture. This is the other side of a paradoxical world in which religious groups
are becoming more assertive as well. It is difficult to predict anything about
the outcome of these antithetical trends.

12

CONCLUSION

GEOPOLITICAL VISIONS: CONTINUITY AND CHANGE

The Second World War has been the most pervasive influence on geopolitical visions in the past half-century. Since war and revolution are the most disturbing events in the life of individuals and nations, this conclusion is not surprising. However, knowing an influence does not imply that one also knows its consequences. Geographical location, history and the amalgamation of both in a national self-concept ultimately determine the effect that worldwide crises or other threats will have on the mind of a people. For certain countries, such as the original European Community members, the effect of the war was to discover a common destiny, while others (such like Argentina and Australia) felt more isolated or more dependent than before, or determined to nip evil in the bud (the United States). In Third World countries – most of them still colonies shortly after the Second World War – the war confirmed an impression already established by Russia's defeat by Japan in 1905: that the European powers were weakened and had lost their moral authority. The world is a system of messages but they are understood differently in different places.

New geopolitical realities can only be interpreted in terms of past experiences. The initial impact of the Second World War was often not a complete overturning of previous visions but rather a confirmation and reinterpretation of prewar positions, even if the result was at odds with the emerging world order. British politicians did not learn that the country could not continue its worldwide imperial obligations: the lesson with most impact was that one should stand firm against future external threats. Argentina learned that the message of the 1929 stock market crash heralding the end of free-market capitalism, was right. Australia sought a new alliance outside its own region to continue its 'white-Australia' ideals. India 'learned' that the exercise of power against the will of a people would be futile. How about the United States, which had entered a qualitatively new situation and responsibility? The Cold War took some years to materialize (as 'theory' would predict) but then the United States did indeed seize upon a new geopolitical vision. Was the effort it took to achieve such a transformation possibly the reason for the domestic purification rite of the witch-hunts in America in the 1950s?

Both self-fulfilling and brand-new visions were challenged after a period of two decades in which the few crises that did occur (like the Korean War) were not really confusing. The war between India and China, the Vietnam War, the development of world trade, and East Asian economic growth, however, were disturbing events for the geopolitical visions of the nations concerned. These events certainly caused practical adaptations in foreign policy but they could not really change the dominant geopolitical vision. The Vietnam War and the war between India and China provide clear examples of such reactions. The Vietnam experience eliminated the American readiness to start head-on military confrontations with 'communist forces' but the American feeling of threat and the vision of territorial loss and gain were maintained. The war with China drove India to seek support from the Soviet Union and to corrupt its own principle of non-alignment, but this idea was kept alive and it was certainly internalized in the 'public' mind.

The resistance to change displayed by a number of national geopolitical visions, in spite of practical challenges, can be explained by the overwhelming impact of the Cold War order in international relations in the four decades after 1945. For Australia, the necessity to integrate in its region was weakened because of the strong anti-communist *cordon sanitaire* in which it participated. The same pertains to the special relationship between Britain and the United States. India, on the other hand, could use its principle of non-alignment to seek freedom and seek advantages from both super-powers. It is not surprising that the fading of Cold War antagonisms since the middle of the 1980s has put pressure on these countries to adopt a new, meaningful vision of external relations. This process evokes problems of identity and hesitations in establishing new foreign policy lines.

The most devastating effects of the current change in geopolitical order have occurred in former Yugoslavia. It is surprising that such effects on the internal stability of countries have remained rather limited so far. Serbia and Russia had to cope not only with new rules of international power but also with the dissolution of their own political order. The geopolitical visions of these countries are challenged more than those of countries in other parts of the world. Some Russian voices suggest picking up the thread broken by the Bolshevik Revolution (that of Eurasianism), albeit with a new interpretation of the role of East Asia. It is too early to judge their impact but the logic is unmistakable.

Like the United States, Germany could not connect immediately after the war with its prewar international situation. Its national identity incurred a 'black-out' from which it is gradually recovering. This means that for Germany a rethinking of the geopolitical situation entails something different from other countries. The Germans might first wish to resume the thread that other countries were able to pick up much earlier. This, at least, is what the unfolding discussion in Germany suggests.

It is impossible to draw general conclusions on the basis of the limited number of country cases which have been explored in the preceding pages, but some observations on change and continuity may be useful in constructing a theoretical framework.

The first conclusion is that in themselves wars, even major wars (like the First World War) or 'lost' wars (like Vietnam) are an insufficient cause for changing geopolitical visions. A change requires a substantial exposure to new security risks (that is, new interactions).[1] My assessment is that this happened with the United States after the Second World War, in spite of its victory, whereas it did not happen after the Vietnam War. However, the latter war and its aftermath, the American withdrawal from Asia, meant an important change for Australia, and here it gave the first impetus with one voice towards a redefinition of national identity and new visions of the region.

The second conclusion is that important shifts in global power (as occurred after the Second World War) hampered the adjustment of the geopolitical visions of minor powers (for example, Britain, Australia, India) to a new regional reality. This apparently occurs when the global order (Cold War) fits in with existing national-identity concepts. Conversely, the current ending of the Cold War order seems to be stimulating some countries (Germany and Russia) to resume discussion on the national geopolitical condition. One wonders whether both changes (to a more regional and to a more national sense of security) can be combined or will produce new tensions in the next century.

Finally, it can be established that the clearest examples (in this book) of either a redefinition of, or attempts to change, geopolitical visions are provided by the United States, Argentina and Australia. It is their drive to become enmeshed in the world (in the case of the United States) or their semi-colonial dependency on a global network that made them liable to the great transformations in the twentieth century: the changing world economy and the decay of European imperial power. Their geographic separation from the (other) centres of power in the world made them prone to visions of forward defence with a wider scope for relocation than the continental security system offers.

THE NATURE OF GEOPOLITICAL VISIONS

In spite of their name, geopolitical visions rarely satisfy the classic realist description of the world as an anarchic universe with the national state as a safe haven. Most geopolitical visions portray the world as divided between two systems, with one's own nation in a strategic position either to defend the cause of right or to act as an intermediary. Countries in a comparatively isolated geographical position, like the United States, Argentina and Australia, have adopted the first view. The Cold War suited their geography. The self-definition as a bridge between East and West is very popular on the Eurasian continent (Germany, Russia and India).

Thus, geopolitical visions in the twentieth century seem to betray a deep longing for a community which transcends national boundaries. However, this does not mean that the national position is made indistinguishable from those of a country's allies. Actually, the wider community is often used for national glorification. In his war with Iran, Saddam Hussein always emphasized the Iraqi position on the Arab frontline. He also exploited diverse symbolic means to express the Iraqi leadership role in the Arab world. In the

Cold War order, the leadership role was of course reserved to the United States, but Britain tried to take over this role, as an extension of American power, in Europe.

One may suppose that the self-image of being an intermediary requires a certain strength or size, which indeed is the case for Germany, Russia and India. Smaller countries, however, have to elaborate a special role for themselves within a wider community. This explains why the East European countries are so anxious to resurrect the European border between East and West, but now in a cultural sense and through NATO membership.[2] They want to be sure to be on the right side. In comparison, the successor states of former Yugoslavia are in an awkward position. Whereas the other East European states define their current situation as a renewal of their repressed but true European identity, the Yugoslavian successors are faced with the loss of a more or less self-chosen identity: non-alignment. Non-alignment (or, being an intermediary) cannot have any meaning for the small nations which have succeeded Yugoslavia, but the problem of a re-defined identity is greatest for Serbia. Serbia, having always defined itself as a frontier state of European civilization, has become the victim of its own desperate attempts to increase its security. The sense of being rejected by Europe in the course of the chain reactions that followed the dissolution of the Yugoslavian state has possibly even reinforced irrational violence. Greece is in a similar position in relation to Europe, but there is fortunately no geopolitical change at the moment to kindle emotions that are just as tense here as in other parts of the Balkans.

Another distinction that can be made between geopolitical visions depends on straightforward physical conditions: that is, between the visions of oceanic states and continental states. In the visions of Britain and the United States, boundaries and relations with neighbours are comparatively unimportant. The geopolitical vision in such countries is a system of connections across the world, rather than a connection with neighbours. Germany, with its multitude of neighbours *and* ethnic affinities across its borders, is the outstanding representative of the other type.

THE PROMINENCE OF FOREIGN AFFAIRS

It is tempting to present geopolitical visions as properties of countries, as if a country could think and see. Nobody would really endorse such a view, but the problem is that foreign policy is sometimes the only, or the most, explicit indicator of the prevailing visions of a country. By relying heavily on foreign policy as a kind of materialized opinion, one essentially views countries as black boxes with hypothetical minds. This is an exaggerated presentation of an intellectual position. Actually, studies of foreign policy visions will always attempt to include individual reflections and opinions that are being expressed outside the decision-making sphere. Public opinion, however, is a problem of its own. Foreign policy is a field with a very low profile in the daily life of most people. If no fundamental crisis occurs, most people do not bother about the outside world.

Domestic problems and events dominate the media's coverage of news and one has usually to turn to the few 'quality' newspapers to find adequate coverage of international events. Small nations have more reason to be aware of the outside world than large ones but, even in the 'quality' papers of small countries, foreign events may take up only 5 per cent of total column-inches devoted to news (as in the Swedish *Svenska Dagbladet* in 1984).[3] Other useful indicators of the importance of foreign affairs to the citizen can be derived from election campaigns. National elections focus on a rather limited set of issues. From inventories of such issues we can deduce that on average 10 to 13 per cent of election issues concern international problems (including national defence and foreign trade).[4] Only a few countries whose foreign relations were severely affected by the aftermath of the Second World War displayed higher figures during the first postwar decades, such as Japan (28 per cent), the United States (18 per cent) and the Federal Republic of Germany (16 per cent). This seems to indicate that 'geopolitical visions' are poorly articulated in the mind of the average citizen and of little importance in daily politics. Domestic affairs are, roughly speaking, ten times as important as foreign affairs.

One of the occasions on which geopolitical visions will be put into words is during an internal conflict on foreign policy aims. The emergence of opposite views, however, also means that the unifying influence of geography and history on visions of the world has its limits. A special case is the conflict between public opinion and the foreign policy-makers of a country. Since geopolitical visions have a strongly moral slant, the practice of statemanship, frequently dominated as it is by pragmatic decisions and compromise, may clash with public geopolitical visions. In such (rare) occasions, the pursuit of a foreign policy may cause a national uproar that provides a sudden insight into the existence and nature of deeply rooted visions of the world. During the first half of the nineteenth century the British Foreign Office had difficulties in countering an increasing Russophobia (inspired by Russian cruelties in Poland), while during the Second World War it had to discourage the public's Russophilia. A surge of public distrust in the Soviet Union, which the American president and foreign secretary were not yet prepared to follow, may have occurred shortly after the war in the United States, if we are to believe the British envoys. Disentanglement from the Vietnam War and the renouncing of subsequent military involvement abroad was enforced by American public opinion (which was the realist party in this instance?). India's 'public' opinion was aroused by the absence of an official response to Soviet imperialism in Hungary, Czechoslovakia and Afghanistan. In spite of its revolutionary 'mandate', the Soviet Union (or Russia) has the dubious honour of having infuriated many peoples (and not only those of its Cold War adversaries). The United States has done the same in Europe and the Third World. Public hostility to the United States occurred in Europe with the demonstrations against the Vietnam War in the 1960s and the peace demonstrations at the beginning of the 1980s. In the Netherlands mass demonstrations of half a million people against the deployment of cruise missiles compelled the Dutch government to adopt a more cautious political line within NATO. The general pattern seems to be that public emotions are particularly provoked by crude power politics, even when this comes from allies. In this period only the American public rose mainly against their own statesmen. Elsewhere the outrage (partly) concerns another malefactor than the home government, a fact that reveals the involvement of a nationalist motivation.

Usually foreign policy-makers will anticipate the mood of the public on crucial subjects. Consequently, the absence of contested foreign policy issues does not prove that political leaders manipulate public opinion. Because decision-makers often share the same national background and place-bound experiences as the public, their aims are often tacitly in tune with public sentiments. However, we should also realize that a large part of foreign policy concerns actions and topics that do not stir the public mind. Even in the case of manipulation, official rhetoric still has to rely on geopolitical images, understood by the majority of the people. The clichés of American containment policy during the Cold War more or less deliberately deceived the public. The most obvious examples of manipulated geopolitical visions are, of course, to be found in undemocratic countries. Saddam Hussein knew very well how to take advantage of Arab anti-Iranian and anti-Kuwait feelings, but his attempts to create a positive Iraqi identity were inconsistent with these visions and were also subverted by his own system of terror.

The other type of disagreement about geopolitical visions is more or less limited to the intellectual sphere. Nonetheless, new visions of a country's elite may very well express rising feelings of discomfort in the masses and portend future policy changes. The philosophers of Eurasianism in Russia and of Middle Europe in Germany are voicing new needs for identity that result from either the effects of a deeply traumatic event (such as the dissolution of the Soviet Union) or the emergence of a new generation lacking the remembrance of such things past (for example, the Second World War).

THE LANDSCAPE OF SECURITY

The definition given in Chapter 1 associates geopolitical visions with feelings concerning security and/or a mission in the world. Traditionally war was the most tangible external security threat. Changing weapon technology after the Second World War has diminished the significance, particularly in western Europe, of frontiers for averting security risks emanating from war. Frontier security gave way to a general feeling of vulnerability to nuclear destruction. This also happened to a certain extent in the United States despite the 'grace' (protection) afforded by its geography. This nuclear fear was significantly less noticeable in Argentina, Australia and India. Europeans, in particular, realized that frontiers would have no relevance when once again their subcontinent became a theatre of war: no type of defence, however strong, would be enough. Prevention of war by means other than military reinforcement became recognized as the only rational option.

Even when frontiers lose their meaning as a protection against war, security feelings may remain attached to the national territory for other reasons. As the postwar British 'invasion' novels show, new fears may become attached to one's identity or culture. Whereas the British (more precisely the English) feared 'continental culture' as the great equalizer, the French took measures to push back foreign (particularly, American) elements in their culture industry. The new German nationalism may be sensitive to the same threats.

The perceived threat to their economic security is another dimension that may currently be much more important to most people than the threat of war. The experience is overwhelmingly non-territorial, and related to the family and to the organizations in an individual's life, but one may ask what territorial framework is activated if a person does appeal to a (quasi-) political authority for economic support or protection. In Europe the state is still the dominant refuge when it comes to political solutions to unemployment. In the 1980s the importance of the state even increased in the public perception of this problem, as a comparative content analysis of newspapers in three countries has revealed.[5] A regional assertiveness emerged at the same time that was partly a result of the unwillingness or inability of central governments to solve specific regional problems.

In certain cases these regions are pursuing what one may call a foreign policy. This concerns either economic co-operation across the national border or an appeal to Europe (the EU) as a means to emphasize the regional cultural identity (of, for example, Scotland or Catalonia). It remains unclear to what extent these attempts are really concerned with individual feelings about security. The political socialization of most citizens is still entwined with their national history and place in the world. Even rewriting the history of Europe as a history of regions, as Christopher Harvie has done, cannot alter that condition.[6] The case studies in this book do not illuminate such regional variants of geopolitical visions, and it must be acknowledged that neither, the heuristic method followed (using international crises and national foreign policy-makers as a core of information) was particularly appropriate for discovering such variants.

In Third World countries the state is still more removed from problems of individual economic security. Village, tribe or clan are the most tangible frontiers of security. Although such ties have been undermined by the process of nationalization and modernization, this type of security arrangement remains, in some form or another, strong enough to be considered as uniquely suited to the post-modern world of crumbling frontiers. An Indian can move to many metropolitan centres in the world and link up with a network of people who are familiar, in a religious or local sense, thus providing a kind of security that an American or European moving out of their country will have to do without.

These observations on the various arrangements to which people resort in pursuing economic security do not imply that geopolitical visions, particularly at the state level, have no economic relevance. On the contrary, new ideas concerning national identity and external relations, such as the Eurasian perspective in Russia and reorientation towards Asia in Australia, have a very strong economic background. They show that, in the West, the state continues to play a pivotal role in the economic field.

A final category of security needs is related to the environment. The state seems particularly ill-equipped to solve the most pressing threats to the global environment. However, as regards control over resources such as rivers, the sea and mineral resources, the state may play a more pervasive role. Pollution across national boundaries has added a completely new threat to the problem of external security. As a result of geographical conditions, problems may even inequitably accumulate in certain countries (for example, in the delta of a great river), and in this way get ingrained in people's experience of the world. There is much evidence concerning the public fear of environmental threats, but this awareness is usually not incorporated in a geopolitical vision, a perspective on the relationship between one's own political unit and other places. Something approximating to this idea is the West European fear of unsafe nuclear reactors in Eastern Europe. That such feelings also have a strong national bias was made clear by the inadequate reactions of the French to the Chernobyl disaster in 1986.

NATIONAL IDENTITY, MISSIONS AND GEOPOLITICAL VISIONS

The case studies in this book have shown that ideas about national identity collide with power structures in the world or with other (geo-political) constraints. Geopolitical visions develop in order to cope with such threats arising from the environment, in order to maintain pride, or just to legitimate aggression. These visions may be considered a compromise between the more ideological and mythic aspects of national identity and the resistant reality. Do people still need myths and missions? Nothing in our case studies shows an increased willingness to do away with national values or aims. In countries like Russia, Germany and Australia, the situation seems even to be the opposite. After a period in which a belief in universal values was propagated, now voices are heard warning that such values may not automatically be serving the national interest. These warnings are without exception coupled with geopolitical visions that are experienced by a counter-elite as very enlightening.

However, as Henry Kissinger said, 'we live in an age of transition.' In this last decade of the century, the geopolitical world order is again changing. It may be that everything will soon transform, that we are looking at the last convulsions of an outdated way of organizing and defining ourselves. One of the outcomes may be the emergence of a limited number of world regions, such as Europe, the Far East, America, etc. As I have said above, this does not necessarily extinguish national distinctions. It may even provide new meanings and new rivalries between states.

Another outcome may simply be the further erosion of the state which seems to occur in the United States and India.[7] This development is coupled with a simultaneous reinforcement of other kinds of group identity. Such groups have a very low awareness of the world outside. Their use of national symbols (the flag, building the holy land) is deceptive. There is no geopolitical vision involved, merely the wish to prize one's own movement as the True Religion.

Whether the state withdraws or is re-emphasized, the reinforcement of identities is often at stake. The new trans-national economic reality seems not to diminish the need for a local or national political discourse. The essential question is, of course, whether the resulting geopolitical visions and discourses will interfere with the new trans-national economic 'reality' or remain just a rhetoric accompaniment that serves essentially to boost pride and to diminish pain. The cases in this book have demonstrated that such visions have not only served narcissistic aims but have also interfered with the long-term interests of a state or group. It requires much optimism to believe that international relations in the future will remain free from the ideological perspectives of particular groups. The end of history has not yet arrived.

NOTES

CHAPTER 1: THE NATIONAL EXPERIENCE OF PLACE

1 *Die Zeit*, 30 March 1990, p. 7.
2 G. Ó Tuathail and J. Agnew, 'Geopolitics and discourse: practical geopolitical reasoning in American foreign policy', *Political Geography*, 1992, vol. 11, pp. 190–204.
3 E.W. Said, *Orientalism*, London, Routledge & Kegan Paul, 1978.
4 K. Cutts Dougherty, M. Eisenhart and P. Webley, 'The role of social representations and national identities in the development of territorial knowledge: a study of political socialization in Argentina and England', *American Educational Research Journal*, 1992, vol. 29, pp. 809–36.
5 R.W. Kates, 'Experiencing the environment as hazard', in S. Wapner, S.B. Cohen and B. Kaplan, *Experiencing the Environment*, New York, Plenum Press, 1976.
6 Geopolitics continued to be practised in Latin America. A rare example of an American writer applying an explicit geopolitical perspective in the early 1960s is S. B. Cohen, *Geography and Politics in a World Divided*, New York, Random House, 1963.
7 L.W. Hepple, 'The revival of geopolitics', *Political Geography Quarterly*, 1986, vol. 5 (suppl.), pp. 21–36.
8 Y. Lacoste, *La géographie, ça sert d'abord à faire la guerre*, Paris, Maspéro, 1976.
9 G. Ó Tuathail, 'The language and nature of the "new geopolitics" – the case of US–El Salvador relations', *Political Geography Quarterly*, 1986, vol. 5, pp. 73–85; M.J. Shapiro, 'The constitution of the Central American other', in M.J. Shapiro, *The Politics of Representation: Writing Practices in Biography, Photography, and Policy Analysis*, Madison, University of Wisconsin Press, 1988; S. Dalby, 'American security discourse: the persistence of geopolitics', *Political Geography Quarterly*, 1990, vol. 9, pp. 171–88.
10 D. Campbell, *Writing Security: United States Foreign Policy and the Politics of Identity*, Manchester, Manchester University Press, 1992.
11 J.M. Welsh, 'The role of the inner enemy in European self-definition: identity, culture and international relations theory', *History of European Ideas*, 1994, vol. 19, pp. 53–61.
12 M. Butor, 'L'Espace du roman', in M. Butor, *Répertoire II*, Paris, Éditions de Minuit, 1964. (Emphasis added in quotation.)
13 C.F. Alger, 'Perceiving, analysing and coping with the local–global nexus', *International Social Science Journal*, 1988, vol. 117, pp. 321–39; C.F. Alger, 'The world relations of cities: closing the gap between social science paradigm and everyday human experience', *International Studies Quarterly*, 1990, vol. 34, pp. 493–518; C.F. Alger, 'Local response to global intrusions', in Z. Mlinar, *Globalization and Territorial Identities*, Aldershot, Avebury, 1992, pp. 77–104.

14 W. Bloom, *Personal Identity, National Identity and International Relations*, Cambridge, Cambridge University Press, 1990.

15 E.W. Said, *Orientalism*, London, Routledge & Kegan Paul, 1978. The 'Orient' in Said's work roughly means Asia, more precisely the Islamic countries and particularly the Arab world (including the North African countries!).

16 K. Makiya, *Cruelty and Silence: War, Tyranny and Uprising in the Arab World*, London, Jonathan Cape, 1993.

17 E.W. Said, 'Afterword to the 1995 printing', in E.W. Said, *Orientalism*, London, Penguin Books, 1995.

18 M. Bassin, 'Russian geographers and the "national mission" in the Far East', in D. Hooson (ed.), *Geography and National Identity*, Oxford, Blackwell, 1994, pp. 112–33. A similar sequence of identifications with a frontier occurred in former Yugoslavia where the Slovenes, the Croats and the Serbs all had their own arguments for representing the last outpost of European culture.

19 H. Varenne, 'The question of European nationalism', in T.M. Wilson and M.E. Smith, *Cultural Change and the New Europe*, Boulder, Westview Press, 1993, p. 235.

20 Smith and Wertman did not find any indication of a wish to weaken the relations with the United States in the future: S.K. Smith and D.A. Wertman, *US–West European Relations during the Reagan Years: The Perspective of West European Publics*, London, Macmillan, 1992.

21 M. Kundera, 'Un occident kidnappé, ou la tragédie de l'Europe centrale', *Le Débat*, 1983, vol. 27, pp. 2–24. English version ('The tragedy of Central Europe') in *New York Review of Books*, 26 April 1984.

22 *Time* magazine, 26 August 1991.

23 Varenne, 'The question', p. 237.

24 R. Strassoldo, 'Globalism and localism: theoretical reflections and some evidence', in Z. Mlinar, *Globalization and Territorial Identities*, pp. 35–60.

25 See C. Tilly, *Coercion, Capital and European States: AD 990–1992*, Cambridge, Blackwell, 1990.

26 Called 'bourgeois-regions' by Christopher Harvie. C. Harvie, *The Rise of Regional Europe*, London, Routledge, 1994.

27 See, for a more extensive discussion with similar conclusions: A.D. Smith, 'National identity and the idea of European unity', *International Affairs*, 1992, vol. 68, pp. 55–76.

28 I base this impression on an unpublished analysis of the Eurobaromètre public opinion files (1970–89; 357,237 cases) and the US General Social Survey file (1972–90; 26,265 cases).

29 Not necessarily the place of residence: compare the Jewish position before the Second World War.

30 A.D. Smith, *National Identity*, London, Penguin Books, 1991, p. 14.

31 W. Wallace, 'Foreign policy and national identity in the United Kingdom', *International Affairs*, 1991, vol. 67, pp. 65–80.

32 J.L. Gaddis, *Strategies of Containment: A Critical Appraisal of Postwar American National Security Policy*, New York, Oxford University Press, 1982, p. ix.

33 T. Varlin, 'La Mort de Che Guevara', *Hérodote*, 1977, vol. 5, pp. 39–81.

34 O.R. Holsti, 'The belief system and national images: a case study', *Journal of Conflict Resolution*, 1962, vol. 6, pp. 244–52. This is not the place to enter into a discussion on this field of studies. I just mention some pioneer works which have helped my own understanding of the subject in the course of time. K. Boulding, *The Image*, Ann Arbor, University of Michigan Press, 1956; D. Lowenthal, 'Geography, experience and imagination: towards a geographical epistemology', *Annals of the Association of American Geographers*, 1961, vol. 51, pp. 241–60; H. Wilensky, *Organizational Intelligence*, New York, Basic Books, 1967; R. Jervis, *Perception and Misperception in International Politics*, Princeton, Princeton University Press, 1976.

35 D.W. Blum, 'The Soviet foreign policy belief system: beliefs, politics and foreign policy outcomes', *International Studies Quarterly*, 1993, vol. 37, pp. 373–94.

36 M. Hauner, *What Is Asia to Us? Russia's Asian Heartland Yesterday and Today*, London, Routledge, 1992 (1st edn 1990), p. 24.

CHAPTER 2: THE COUNTRY OF *ANGST*
(Germany)

1 F. Ratzel, *Deutschland. Einführung in die Heimatkunde*, Leipzig, Fr. Wilhelm Grunow, 1898.
2 H. Schmidt, *Menschen und Mächte*, Berlin, Siedler Verlag, 1987, p. 450.
3 J. Thies, 'Perspektiven deutscher Aussenpolitik', in R. Zitelmann, K. Weissmann and M. Grossheim, *Westbindung. Chancen und Risiken für Deutschland*, Frankfurt am Main, Propyläen, 1993, p. 528.
4 Perhaps Pearl Harbor is an exception.
5 W.R. Mead, 'The once and future Reich', *World Policy Journal*, 1990, vol. 7, p. 607.
6 A perspective which is also adhered to by some non-German historians like David Calleo: D.P. Calleo, *The German Problem Reconsidered: Germany and the World Order, 1870 to the Present*, Cambridge, Cambridge University Press, 1978.
7 '. . . wesentliche und kontinuierliche sozialpsychologische Grundbefindlichkeit in Deutschland'. Quoted in W.D. Gruner, *Die deutsche Frage. Ein Problem der europäischen Geschichte seit 1800*, Munich, C.H. Beck, 1985, p. 112.
8 H. Sprout and M. Sprout, 'Environmental factors in the study of international politics', *Journal of Conflict Resolution*, 1957, vol. 1, pp. 309–28.
9 An idea derived from Clausewitz.
10 At least among nineteenth-century German intellectuals and artists.
11 With an edition of 50 million copies sold at the beginning of the 1980s, Karl May is the most widely read German writer.
12 Klaus Mann talks of the blend of impertinence, quotations from the Bible and descriptions of the organizing of massacres, which must have had an irresistible attraction for sick minds like that of the Führer. K. Mann, *Distinguished Visitors. Der Amerikanische Traum*, Munich, Edition Spangenberg, 1992 (1939).
13 A.J.P. Taylor, *The Course of German History*, London, Methuen, 1978 (1st edn 1945).
14 *Frankfurter Zeitung*, 17 June 1899. H. Wollschläger, *Karl May in Selbstzeugnisse und Bilddokumente*, Reinbek/Hamburg, Rowohlt, 1965.
15 May adopts the English term 'greenhorn'. J. Schulte-Sasse, 'Karl Mays Amerika-Exotik und deutsche Wirklichkeit. Zur sozialpsychologischen Funktion von Trivialliteratur im wilhelminischen Deutschland', in H. Schmiedt, *Karl May*, Frankfurt am Main, Suhrkamp, 1983, pp. 101–29.
16 Ironically, the empiricist Anglo-Saxons had their own hero with the same qualities at his disposal: Conan Doyle's Sherlock Holmes.
17 H. James, *A German Identity: 1770 to the Present Day*, London, Phoenix, 1989/1994.
18 M. Swales, 'The problem of nineteenth-century German realism', in N. Boyle and M. Swales, *Realism in European Literature: Essays in Honour of J.P. Stern*, Cambridge, Cambridge University Press, 1986.
19 Quoted in M. Eksteins, *Rites of Spring: The Great War and the Birth of the Modern Age*, New York, Doubleday, 1989, pp. 68–74.
20 James, *A German Identity*, p. 106.
21 J. Schulte-Sasse, 'Karl Mays Amerika-Exotik', pp. 107ff.
22 C. Wagner, *Die Tagebücher (I)*, M. Gregor-Dellin and D. Mack (eds), Munich/Zurich, R. Piper, 1976, p. 259.
23 Already in 1849 Friedrich Engels had described the East European nations as counter-revolutionary peoples that would be exterminated in a coming world war (!): H. James, *A German Identity*, p. 53.
24 W.Wippermann, *Der 'Deutsche Drang nach Osten'. Ideologie und Wirklichkeit eines*

politischen Schlagwortes, Darmstadt, Wissenschaftliche Buchgesellschaft, 1981, pp. 92ff.

25 M. Burleigh, 'Scholarship, state and nation 1918–1945', in J. Breuilly (ed.), *The State of Germany: The National Idea in the Making, Unmaking and Remaking of a Modern Nation-State*, London, Longman, 1992; M. Rössler, '*Wissenschaft und Lebensraum*'. Geographische Ostforschung im Nationalsozialismus, Berlin/Hamburg, Dietrich Reimer Verlag, 1990.

26 Quoted in K. Filipp, *Germany Sublime and German Sublimations*, Munster, Waxmann, 1993, p. 93.

27 Rössler, *Wissenschaft und Lebensraum*, p. 57.

28 See Burleigh, 'Scholarship, state and nation', p. 133. One might speculate on the question of whether Braudel's message is equally tendentious.

29 An effect in the central nervous system suggesting that a lost limb is still in place.

30 Rössler, *Wissenschaft und Lebensraum*, p. 149. National Socialist research managers were critical of Christaller. They found his work too formalistic and insufficiently 'political-organic'.

31 G. Bakker, *Duitse Geopolitiek 1919–1945: een Imperialistische Ideologie*, Assen, Van Gorcum, 1967. Germany was undoubtedly one of the most densely populated countries in Europe, but as a result of this manipulation the contrasts became very extreme.

32 Rössler, *Wissenschaft und Lebensraum*, p. 172.

33 Of course, they did not agree with the western border, which had to be shifted to the Vosges and Ardennes ridges. Concepts of *Mitteleuropa* (often a cloak for plans about a future Germany) changed a lot in the course of time. In the first part of the nineteenth century it always included France and the Low Countries. After the consolidation of the Bismarck Reich, politically engaged writers excluded France and focused more on the territories to the east and south-east. This was opposed by 'pure' geographers, who would not permit the use of political criteria. After the First World War, geographers became less scrupulous: H.-D. Schultz, 'Fantasies of Mitte: *Mittellage* and *Mitteleuropa* in German geographical discussion in the 19th and 20th centuries', *Political Geography Quarterly*, 1989, vol. 8, pp. 315–39. As I have indicated, Ratzel accepted the political argument for the independence of the Netherlands. Nevertheless, he saw a more profound geographical message in that the Dutch would once 'acknowledge that in the current world situation their existence as a separate nation will, at least from an economic point of view, not be fitting any more'. Ratzel, *Deutschland*, p. 288.

34 Bakker, *Duitse Geopolitiek*, pp. 105ff. It will be clear that there was no room for an independent Poland, in whatever form, in this conception.

35 Bakker, *Duitse Geopolitiek*, p. 120.

36 E. Schulz, *Die deutsche Nation in Europa. Internationale und historische Dimensionen*, Bonn, Europa Union Verlag, 1982, pp. 240–1.

37 Konrad Adenauer and other conservatives were even counteracting attempts to change foreign policy favouring reunification in the 1950s. J. Ossenbrügge, 'Territorial ideologies in West Germany 1945–1985: between geopolitics and regionalist attitudes', *Political Geography Quarterly*, 1989, vol. 8, p. 391.

38 Schulz, *Die deutsche Nation*, p. 243.

39 T. Garton Ash, *In Europe's Name: Germany and the Divided Continent*, New York, Random House, 1993.

40 Schmidt, *Menschen und Mächte*, p. 304.

41 Garton Ash, *In Europe's Name*, p. 290.

42 E.S. Herman and N. Chomsky, *Manufacturing Consent: The Political Economy of the Mass Media*, New York, Vintage, 1988.

43 Garton Ash, *In Europe's Name*, p. 367.

44 R. Zitelmann, K. Weissmann and M. Grossheim, *Westbindung. Chancen und Risiken für Deutschland*, Frankfurt am Main, Propyläen Verlag, 1993.

45 K.R. Röhl, 'Morgenthau und Antifa. Über den Selbsthass der Deutschen', in H.

Schwilk and U. Schacht (eds), *Die selbstbewusste Nation. 'Anschwellender Bocksgesang' und weitere Beiträge zu einer deutschen Debatte*, Berlin, Ullstein, 1994,' pp. 96–7.

46 A. Mechtersheimer, 'Nation und Internationalismus. Über nationales Selbstbewusstsein als Bedingung des Friedens', in H. Schwilk and U. Schacht (eds), *Die selbstbewusste Nation*, p. 349.

47 Because they wish to conceive of Germany as a normal country.

48 A.C. Wanders, G. Nicolas and G. Parker, *Géovision allemande des Europes*, Lausanne, Eratosthène, 1995.

49 K.-E. Hahn, 'Westbindung und Interessenlage. Über die Renaissance der Geopolitik', in H. Schwilk and U. Schacht (eds), *Die selbstbewusste Nation*, pp. 327–44.

50 For example, the German writer Botho Strauss, whose 'Anschwellender Bocksgesang' (Swelling song of the goats) provided the motto for the collection of essays in H. Schwilk and U. Schacht (eds), *Die selbstbewusste Nation*.

51 M. Mertes, 'Germany's social and political culture: change through consensus?', *Daedalus*, 1994, vol. 123, no. 1, pp. 1–32.

CHAPTER 3: ABSENT BECAUSE OF EMPIRE
(Britain)

1 'They are brave soldiers these English volunteers.' 'Yes, yes . . . they run so well.' 'Oh yes, but they are not quite so fast as the French rabble.' 'Certainly, and they make rather good riflemen here.' 'You are right, tall Peter. If the bastards had been drilled as well as they can shoot, we wouldn't be here.' *The Battle of Dorking: Reminiscences of a Volunteer*, Anonymous (G.T. Chesney), Edinburgh, William Blackwood & Sons, 1871, pp. 57–8.

2 It did not even mean the start of invasion literature as such. However, earlier writings (from 1763 on) were pamphlets, poems or stories which had substantially less impact than *The Battle of Dorking*.

3 L. Colley, *Britons: Forging the Nation 1707–1837*, London, Pimlico, 1994 (1st edn 1992), p. 306.

4 M. Spiering, *Englishness: Foreigners and Images of National Identity in Postwar Literature*, Amsterdam, PhD thesis, University of Amsterdam, 1993, pp. 125–6.

5 Spiering, *Englishness*, p. 128.

6 P. Anderson, 'Nation-states and national identity', *London Review of Books*, 9 May 1991.

7 B. Porter, *Britain, Europe and the World 1850–1986: Delusions of Grandeur*, London, Allen & Unwin, 1986.

8 Porter, *Britain*, pp. 36, 71, etc.

9 C.J. Bartlett, *British Foreign Policy in the Twentieth Century*, Basingstoke, Macmillan, 1989, p. 1.

10 J.H. Gleason, *The Genesis of Russophobia in Great Britain: A Study of the Interaction of Policy and Opinion*, Cambridge, Harvard University Press, 1950, p. 50.

11 M. Hauner, *What Is Asia to Us?*, Routledge, London, 1992 (1st edn 1990), p. 80.

12 P.J. Taylor, *Britain and the Cold War: 1945 as a Geopolitical Transition*, London, Pinter, 1990, pp. 12, 136ff.

13 Taylor, *Britain and the Cold War*, p. 110.

14 A much earlier statement than the celebrated telegram of George Kennan. A.M. Schlesinger Jr, 'Why the Cold War?', in A.M. Schlesinger Jr, *The Cycles of American History*, Boston, Houghton Mifflin, 1986, p. 181.

15 P.G. Boyle, 'The British foreign office view of Soviet–American relations, 1945–46', *Diplomatic History*, 1979, vol. 3, pp. 307–20.

16 Taylor, *Britain and the Cold War*, p. 55.

17 P.M.H. Bell, *John Bull and the Bear: Public Opinion, Foreign Policy and the Soviet Union 1941–1945*, London, Edward Arnold, 1990, pp. 103ff.

18 Bell, *John Bull*, p. 188.
19 Boyle, 'The British foreign office', p. 311.
20 Boyle, 'The British foreign office', p. 311.
21 Boyle, 'The British foreign office', pp. 313–14.
22 Bartlett, *British Foreign Policy*, p. 85. The continuous emphasis on the 'Irish dimension' in each possible security threat will not be discussed here. Ireland was never considered as a threat on its own but rather as a strategic asset 'to be kept at all hazards'. See M. McKinley, 'To be "kept at all hazards": the British construction, and spatio-temporal deconstruction of Ireland as a subdued threat and strategic asset', *History of European Ideas*, 1994, vol. 19, pp. 101–13.
23 W. Wallace, 'Foreign policy and national identity in the United Kingdom', *International Affairs*, 1991, vol. 67, pp. 65–80.
24 M. Blackwell, *Clinging to Grandeur: British Attitudes and Foreign Policy in the Aftermath of the Second World War*, Westport, Greenwood Press, 1993, p. 52.
25 According to an analysis of school history textbooks from before 1914 which was published in 1929. Blackwell, *Clinging to Grandeur*, p. 51.
26 Blackwell, *Clinging to Grandeur*, p. 72.
27 Blackwell, *Clinging to Grandeur*, p. 148.
28 Blackwell, *Clinging to Grandeur*, p. 162.
29 Wallace, 'Foreign policy and national identity'. For the continuity of the special relationship, see also: D. Sanders and G. Edwards, 'Consensus and diversity in elite opinion: the views of the British foreign policy elite in the early 1990s', *Political Studies*, 1994, vol. 42, pp. 413–40.
30 W. Wallace, 'British foreign policy after the Cold War', *International Affairs*, 1992, vol. 68, p. 429.
31 Wallace, 'British foreign policy', p. 441.
32 D. Miller, 'Reflections on British national identity', *New Community*, 1995, vol. 21, p. 157.
32 D. Marquand, 'After Whig imperialism: can there be a new British identity?', *New Community*, 1995, vol. 21, p. 188.

CHAPTER 4: THE MARCH OF CIVILIZATION: DESTINY AND DOUBTS (United States)

1 J.L. Garreau, *The Nine Nations of North America*, New York, Avon Books, 1982, p. 122.
2 P. Avrich, *The Haymarket Tragedy*, Princeton, Princeton University Press, 1984, p. 215.
3 Avrich, *The Haymarket Tragedy*, p. 218.
4 Although there are some indications that it was an anarchist named George Meng. See P. Avrich, 'The bomb-thrower: a new candidate', in D. Roediger and F. Rosemont (eds), *Haymarket Scrapbook*, Chicago, C.H. Kerr, 1986, pp. 71–3.
5 C. Ashbaugh, 'Judge Gary vs. the people: conspiracy trials in America', in D. Roediger and F. Rosemont, *Haymarket Scrapbook*, pp. 243–6.
6 D. Campbell, *Writing Security*, Manchester, Manchester University Press, 1992, p. 175.
7 G. Wills, 'The new revolutionaries', *New York Review of Books*, 10 August 1995.
8 So-called 'post-structuralists', see K.-J. Dodds, 'Geopolitics and foreign policy: recent developments in Anglo-American political geography and international relations', *Progress in Human Geography*, 1994, vol. 18, p. 192. See, for the American application, D. Campbell, *Writing Security*.
9 R. Slotkin, *The Fatal Environment*, New York, Atheneum, 1985, pp. 301–6.
10 A.M. Schlesinger Jr, *The Disuniting of America*, New York, W.W. Norton, 1992, p. 25

11 G. Ó Tuathail and J. Agnew, 'Geopolitics and discourse', *Political Geography*, 1992, vol. 11, pp. 190–204.
12 Slotkin believes that the positive connotation of the 'war of extermination' is unique to American culture, although 'we have not in actual fact carried through any of the larger genocidal threats implicit in the myth': R. Slotkin, *The Fatal Environment*, p. 62. Elsewhere in this book (p. 26) I have referred to the Prussian historian Heinrich von Treitschke who, writing much in the same vein, boasted of the violence of the German colonization to the east in medieval times by calling it *Völkermord* (genocide). Whereas the historical prototype may have been closer to genocide in the US than in the German case, its 'fulfilment' in the twentieth century reversed this relation.
13 J. Hellmann, *American Myth and the Legacy of Vietnam*, New York, Columbia University Press, 1986, pp. 46–7.
14 F. FitzGerald, *Fire in the Lake*, Boston, Little Brown, 1972, p. 8.
15 C.J. Rooney Jr, *Dreams and Vision: A Study of American Utopias, 1865–1917*, Westport, Greenwood Press, 1985, pp. 4–5.
16 R.W. Tucker and D.C. Hendrickson, *The Imperial Temptation*, New York, Council on Foreign Relations Press, 1992, p. 179.
17 Campbell, *Writing Security*, pp. 157, 153.
18 R.D. Brunner, 'Myth and American politics', *Policy Sciences*, 1994, vol. 27, pp. 1–18.
19 At least in a political sense. See Tucker and Hendrickson, *The Imperial Temptation*.
20 R. Hughes, *Culture of Complaint*, New York, Warner Books, 1994.
21 F.L. Klingberg, 'Cyclical trends in American foreign policy moods and their policy implications', in C.W. Kegley Jr and P.J. McGowan (eds), *Challenges to America*, Beverly Hills, Sage, 1979, pp. 37–55; A.M. Schlesinger Jr, 'The cycles of American politics', in A.M. Schlesinger Jr, *The Cycles of American History*, Boston, Houghton Mifflin, 1986, pp. 23–48.
22 Empirical evidence on political (value) cycles is not very convincing. See: M. Eisner, 'Long-term dynamics of political values in international perspective', *European Journal of Political Research*, 1990, vol. 18, pp. 605–21.
23 H. Kissinger, 'We live in an age of transition', *Daedalus*, 1995, vol. 124, no. 3, p. 101.

CHAPTER 5: THE LAST FRONTIER
(United States)

1 A.M Schlesinger Jr, 'Foreign policy and the American character', in A.M. Schlesinger Jr, *The Cycles of American History*, Boston, Houghton Mifflin, 1986, p. 51.
2 Schlesinger Jr, 'Foreign policy', p. 7.
3 H. Kissinger, 'We live in an age of transition', *Daedalus*, 1995, vol. 124, no. 3, p. 99.
4 Although one might attribute the Iranian crisis to the weakening of the Shah's position as a consequence of the human rights policy of President Carter.
5 D.P. Moynihan, *Pandaemonium*, Oxford, Oxford University Press, 1993, p. 166. The famous secret speech of Gorbachev to a gathering of artists and intellectuals (1986), in which he asked for their special support for his coming attempts to change Soviet society, appeared in American journals (such as *Time*) several months after it had been published in Europe. This is another indication of the inability of American officials and public to grasp the importance of imminent changes in Soviet communism.
6 J. Nijman and H. van der Wusten, 'Breaking the Cold War mould in Europe', in J. O'Loughlin and H. van der Wusten (eds), *The New Political Geography of Eastern Europe*, London, Belhaven Press, 1993.
7 G.R. Sloan, *Geopolitics in United States Strategic Policy, 1890–1987*, Brighton,

Wheatsheaf, 1988, p. 202. See also S. Dalby, 'American security discourse: the persistence of geopolitics', *Political Geography Quarterly*, 1990, vol. 9, pp. 171–88.

8 Although Mackinderian in spirit, the term 'rimland' stems from N.J. Spykman, a geopolitical writer with some influence on President F.D. Roosevelt. See: N.J. Spykman, 'Geography and foreign policy I', *American Political Science Review*, 1938, vol. 32. For a comment on Roosevelt's reservations toward geopolitics see J.L. Harper, *American Visions of Europe*, Cambridge, Cambridge University Press, 1994, p. 40.

9 D. Eisenhower, 7 April 1954. Quoted in Sloan, *Geopolitics*.

10 Revolutions are another major experience. R. Jervis, *Perception and Misperception in International Politics*, Princeton, Princeton University Press, 1976, p. 266.

11 The idea was developed by Russell Weigley in *The American Way of War* (1973). Referenced in R. Slotkin, *The Fatal Environment*, New York, Atheneum, 1985, pp. 303–4.

12 J. Hellmann, *American Myth and the Legacy of Vietnam*, New York, Columbia University Press, 1986, p. 4.

13 In 1966 Nixon still professed his confidence in the domino theory. As Theodore Draper recounts Nixon's words: 'If Vietnam fell, he prophesied, Laos, Cambodia, Thailand, Burma and Indonesia had to be written off; we would have to fight a major war to save the Philippines, the Pacific would become a "red sea", and the United States would have to face up to Chinese Communist aggression as far as Australia in only four or five years.' T. Draper, 'McNamara's peace', *New York Review of Books*, 11 May 1995 (note 1).

14 R.A. Melanson, *Reconstructing Consensus*, New York, St Martin's Press, 1991, p. 61.

15 H. Kissinger, *The White House Years*, London, Weidenfeld & Nicolson/Michael Joseph, 1979, pp. 914–15.

16 This author's phrasing.

17 H.A. Kissinger, 'Balance of power sustained', in G. Allison and G.F. Treverton, *Rethinking America's Security*, New York, W.W. Norton, 1992.

18 Melanson, *Reconstructing Consensus*, p. 58.

19 J. O'Loughlin and R. Grant, 'The political geography of presidential speeches 1946–1987', *Annals of the Association of American Geographers*, 1990, vol. 80, pp. 504–30.

20 Sloan, *Geopolitics*, pp. 168–87.

21 Melanson, *Reconstructing Consensus*, p. 153.

22 N. Chomsky, 'Visions of righteousness', in J.C. Rowe and R. Berg, *The Vietnam War and American Culture*, New York, Columbia University Press, 1991.

23 H.A. Kissinger (ed.), *Report of the National Bipartisan Commission on Central America*, Washington, U.S. Government Printing Office, 1984, p. 7.

24 Kissinger, *Report*, p. 38.

25 L. Schoultz, *National Security and United States Policy Toward Latin America*, Princeton, Princeton University Press, 1987, p. 282. See also M.J. Shapiro, 'The constitution of the Central American other: the case of Guatemala', in, M.J. Shapiro, *The Politics of Representation*, Madison, University of Wisconsin Press, 1987.

26 Kissinger, *Report*, p. 93.

27 FitzGerald, *Fire in the Lake*.

28 D. Lowenthal, 'The American scene', *Geographical Review*, 1968, vol. 58, p. 61–88

29 In what way Vietnamese culture has, in the course of history, moved even farther from its source (China) and its American antipode is not specified by FitzGerald, unless she means the utter inability of Vietnamese culture to change and to assimilate foreign influences.

30 R. Berg and J.C.Rowe, 'The Vietnam war and American memory', in J.C. Rowe and R. Berg, *The Vietnam War and American Culture*, 1991.

31 J.E. Mueller, *War, Presidents and Public Opinion*, New York, University Press of America, 1973.

32 E.R. Wittkopf, *Faces of Internationalism: Public Opinion and American Foreign Policy*, Durham, Duke University Press, 1990, pp. 6–7.
33 Wittkopf, *Faces of Internationalism*, pp. 214–15.
34 J. Atherton, 'The vocabulary of the Vietnam War', in J.-R. Rougé, *L'Opinion américaine devant la guerre du Vietnam*, Paris, Presses de l'Université de Paris-Sorbonne, 1991, p. 158.
35 Hellmann, *American Myth*, pp. 206–7.
36 J. Newman and M. Unsworth, *Future War Novels: An Annotated Bibliography of Works in English Published since 1946*, Phoenix, Oryx Press, 1984. The collection is far from complete, as revealed by I.F. Clarke's study *Voices Prophesying War*, Oxford, Oxford University Press, 2nd edition 1992, which lists about 620 works for the same period. However, some of these (like Golding's *Lord of the Flies*) do not quite comply with the criterion of future war novel, while some others are non-English. Newman and Unsworth's bibliography has the advantage of providing abstracts for each novel. The crude content analysis performed here was only possible on the basis of these abstracts.
37 See Chapter 3 on Britain.
38 Clarke, *Voices*, pp. 215–16.
39 20 January 1989. *Public Papers of the American Presidents: George Bush*.
40 R.D. Brunner, 'Myth and American politics', *Policy Sciences*, 1994, vol. 27, pp. 1–18.
41 O.R. Holsti and J.N. Rosenau, 'The structure of foreign policy beliefs among American opinion leaders – after the Cold War', *Millennium*, 1993, vol. 22, pp. 235–78.
42 Kissinger, 'We live in an age of transition', p. 101.

CHAPTER 6: PERIPHERAL DIGNITY AND PAIN
(Argentina)

1 H.J. Mackinder, 'The geographical pivot of history', *Geographical Journal*, 1904, vol. 23, pp. 421–44.
2 R. Nijinsky, *Nijinsky*, London, Victor Gollancz, 1933.
3 J.R. Scobie, *Buenos Aires: Plaza to Suburb, 1870–1910*, New York, Oxford University Press, 1974, p. 235.
4 Thirty German military advisors came to Argentina and 150–175 Argentine army officers went to Germany for instruction. J.D. Rudolph (ed.), *Argentina: A Country Study*, Washington DC, Headquarters Department of the Army, 1986, p. 288.
5 Likewise in Chile. See J. Child, *Geopolitics and Conflict in South America*, New York, Praeger, 1985.
6 Based on a statement of the Argentine writer Carlos Mastronardi. C.T. Leland, *The Last Happy Men: The Generation of 1922, Fiction and the Argentine Reality*, Syracuse, Syracuse University Press, 1986.
7 The comparison with Australia has often been made. See for example T. Duncan and J. Fogarty, *Australia and Argentina: On Parallel Paths*, Melbourne, Melbourne University Press, 1984; D.C.M. Platt and G. di Tella, *Argentina, Australia and Canada: Studies in Comparative Development 1870–1965*, Oxford, Macmillan, 1986.
8 V.S. Naipaul, *The Return of Eva Peron: With the Killings in Trinidad*, London, André Deutsch, 1980.
9 Lucas Ayarragaray quoted in C.H. Waisman, *Reversal of Development in Argentina: Postwar Counterrevolutionary Policies and their Structural Consequences*, Princeton, Princeton University Press, 1987, p. 45.
10 Waisman, *Reversal of Development*.
11 Domingo Faustino Sarmiento wrote *Facundo: Civilización y barbarie en las pampas argentinas* (1845), in which the *caudillo* is identified as a force in Argentine history and ways are suggested to break with that tradition: general education, a constitution and a liberal economy.

12 Quoted in J.A. Lanús, *La Causa argentina*, Buenos Aires, Emecé Editores, 1988, p. 276.

13 Mafud, *Psicología de la viveza criolla: Contribuciones para una interpretación de la realidad social argentina y americana*, Buenos Aires, Editorial Americalee, 1965.

14 See for example C.T. Leland, *The Last Happy Men*, p. 157; Lanús, *La Causa argentina*, p. 270; D.W. Foster, *The Argentine Generation of 1880: Ideology and Cultural Texts*, Columbia, University of Missouri Press, 1990.

15 Lanús, *La Causa argentina*, p. 264. See also C.A. Loprete, *El Ensueño argentino*, Buenos Aires, Editorial Plus Ultra, 1985, p. 156.

16 Quoted in J.A. Lanús, *La Causa argentina*, Buenos Aires, Emecé Editores, 1988, pp. 316–17.

17 G. Wynia, *Argentina: Illusions and Realities*, New York, Holmes & Meier, 1986, p. 36.

18 Perón himself has always given contradictory explanations of his stay in Italy. In any case it was not a political mission. It is certain that he served some time in a division of the Italian army in the Alps. R. Crassweller, *Perón and the Enigmas of Argentina*, New York, W.W. Norton, 1987, p. 86.

19 D. Rock, *Authoritarian Argentina: The Nationalist Movement, its History and its Impact*, Berkeley, University of California Press, 1993, p. 111.

20 From the defamatory pamphlet by Benjamín Villafañe, *Hora obscura*, 1935. Quoted in Waisman, *Reversal of Development*, p. 220.

21 Waisman, *Reversal of Development*, p. 134.

22 Crassweller, *Perón*, p. 88.

23 Wynia, *Argentina*, p. 73.

24 Naipaul, *The Return of Eva Perón*, p. 95.

25 In the novel *Evarista Carriego*, quoted in Lanús, *La Causa argentina*, p. 286.

26 Loprete states that Argentinians like to believe in myths, magic and 'tall stories' but that the Argentinian mentality has produced few imaginative stories and myths. The geographical vastness and the lack variety in the country would not have stimulated the fantasy. Loprete, *El Ensueño argentino*, p. 80.

27 D. Ferraro, *La Personalidad cultural argentina*, Buenos Aires, Editora de la Palmera, 1985.

28 General Villegas, quoted in Child, *Geopolitics and Conflict in South America*, p. 42.

29 The initial offer of President Menem to participate in the international force in the Gulf seemed a revolutionary break in Argentine history, but it soon appeared that such a move was still impossible.

30 Rock, *Authoritarian Argentina*, p. 108.

31 Rock, *Authoritarian Argentina*, p. 231.

32 Loprete also mentions the Argentinian flight from national disorder through an emphatic interest in the problems in other countries such as the slums of Chicago, race problems in the United States, etc. Loprete, *El Ensueño argentino*, p. 80.

33 This did not prevent the development of a special trade relationship with the Soviet Union.

34 K.-J. Dodds, 'Creating a strategic crisis out of a communist drama? Argentine and South African geo-graphs of the South Atlantic', *European Review of Latin American and Caribbean Studies*, 1994, vol. 56, pp. 33–54.

35 J.D. Carasales, *National Security Concepts of States: Argentina*, New York, UNIDIR/United Nations, 1992.

36 Child, *Geopolitics and Conflict in South America*, p. 78.

37 J.E. Corradi, 'Menem's Argentina, Act II', *Current History*, 1995, vol. 94, pp. 77, 79.

CHAPTER 7: WANDERING IN CIRCLES
(Australia)

1 This text has been previously published in Dutch in *Geografie*, 1994, vol. 3, pp. 25–8. Some new references have been added.

2 J.B. Millar, *Australia in Peace and War*, Botany, Australian National University Press, 1991 (2nd edn).

3 M. Gurry, 'Identifying Australia's region: from Evatt to Evans', *Australian Journal of International Affairs*, 1995, vol. 49, p. 30.

4 B. Appleyard, 'The dinkum Aussie? Strewth?', *Independent*, 8 September 1993.

5 In a review of Manning Clark's *History of Australia*, Paul Johnson calls his work 'an inspired condensation of an inspirational work of historical myth-creation'. He further comments: 'It is not easy for Australian historians to make their country's story exciting . . . but [Clark's imagination] turned Australia's placid history into a thrill-packed adventure story.' P. Johnson, 'Blood on the wattle', *The Times Literary Supplement*, 13 May 1994, p.25.

6 H. Collins, 'Political ideology in Australia: the distinctiveness of a Benthamite society', *Daedalus*, 1985, vol. 114, no. 1, pp. 147–69.

7 J. Colmer, *Australian Autobiography; The Personal Quest*, Melbourne, Oxford University Press, 1989.

8 J. King, *Waltzing Materialism – and Other Attitudes That Have Shaped Australia 1788–1978*, Sidney, Harper & Row, 1978.

9 This attitude scale has become well-known through the cross-national research project (particularly on European data) of Ronald Inglehart. R. Inglehart: *Culture Shift in Advanced Industrial Society*, Princeton, Princeton University Press, 1990. The scales used by Papadakis in Australia are not the same but they can be calibrated (using Inglehart's results as norm) because the test was also administered in some European countries: E. Papadakis, 'Does the new politics have a future?', in F.G. Castles (ed.), *Australia Compared; Peoples: Policies and Politics*, North Sydney, Allen & Unwin, 1991, pp. 239–57.

10 C. Bean, 'Are Australian attitudes to government different? A comparison with five other nations', in Castles (ed.), *Australia Compared*, pp. 74–100.

11 Quoted in N. Jose, 'Cultural identity: "I think I'm something else"', *Daedalus, Journal of the American Academy of Arts and Sciences*, 1985, vol. 114, no. 1, p. 325.

12 P. Carter, *Living in a New Country: History, Travelling and Language*, London, Faber & Faber, 1992.

13 J.L. Richardson, *The Gulf War and Australian Political Culture*, Canberra, Australian National University, 1992, p. 6.

14 Millar, *Australia in Peace and War*, p. 177.

15 Richardson, *The Gulf War*, p. 19.

16 K. Perkins, *Menzies: Last of the Queen's Men*, 1969. Quoted in Gurry, 'Identifying Australia's region', p. 21.

17 N. Meaney, 'The end of "White Australia" and Australia's changing perceptions of Asia, 1945–1990', *Australian Journal of International Affairs*, 1995, vol. 49, p. 180.

18 The other explanation is undoubtedly that Australia, in terms of population, is simply a small country.

19 B. Hawke, *The Hawke Memoirs*, London, Heinemann, 1994. Such revelations have caused some astonishment in people who personally knew Hawke. See K. Edwards, 'A ripping give-'em-hell farewell to politics', *Time*, 29 August 1994.

21 R. Leaver, 'Biting the dust: the imperial conventions within republican pretences', *Australian Journal of Political Science*, 1993, vol. 28, pp. 146–61.

22 Collins, 'Political ideology in Australia'.

23 A. Renouf, *The Frightened Country*, Melbourne, Macmillan, 1979.

24 R. Higgott, 'The politics of Australia's international economic relations: adjustment and two level games', *Australian Journal of Political Science*, 1991, vol. 26, p. 8.

25 A.F. Cooper, 'Between fragmentation and integration: the evolving security discourse in Australia and Canada', *Australian Journal of International Affairs*, 1995, vol. 49, p. 63.

26 One may, for example, wish to protect certain forms of traditional agriculture because they are found useful from an environmental or tourist point of view or simply as a meaningful way to assign social security benefits.

CHAPTER 8: THE EURASIAN DILEMMA
(Russia)

1 I rely on George Kennan's review, 'Autopsy on an empire: the American ambassador's account of the collapse of the Soviet Union', *New York Review of Books*, 16 November 1995, for this rendition of the conclusions.

2 As critical historians may observe, a possible pre-revolutionary period of change is ignored in this model.

3 A.H. Miller, V.L. Hesli and W.M. Reisinger, 'Comparing citizen and elite belief systems in post-Soviet Russia and Ukraine', *Public Opinion Quarterly*, 1995, vol. 59, pp. 1–40.

4 M. Urban, 'The politics of identity in Russia's postcommunist transition: the nation against itself', *Slavic Review*, 1994, vol. 53, p. 735.

5 M. Hauner, *What Is Asia to Us?*, London, Routledge, 1992, p. 12. The quote is from one of Martin Walker's reports from Moscow in the *Guardian* (*Manchester Guardian Weekly*, 14 February 1988).

6 D. Kerr, 'The new Eurasianism: the rise of geopolitics in Russia's foreign policy', *Europe–Asia Studies*, 1995, vol. 47, p. 986.

7 *Current Digest of the Post-Soviet Press*, 1995, vol. 47, 8 March. In 1994 the question was what ideas could unite society.

8 A. Kappeler, 'Some remarks on Russian national identities (sixteenth to nineteenth centuries)', *Ethnic Groups*, 1993, vol. 10, pp. 147–55. M. Raeff, 'The people, the intelligentsia and Russian political culture', *Political Studies*, 1993, vol. 41, pp. 93–106.

9 J. Borodaj and A. Nikiforov, 'Between East and West: Russian renewal and the future', *Studies in East European Thought*, 1995, vol. 47, p. 107.

10 M. Bassin, 'Russian geographers and the "national mission" in the Far East', in D. Hooson (ed.), *Geography and National Identity*, Oxford, Blackwell, 1994, p. 117.

11 V. Okhrimenko, 'Jirinovski et "le bond final de la Russie sur le Sud"', *Hérodote*, 1993, vol. 72/73, pp. 119–27.

12 In 1992, 37 per cent of the Russian mass respondents said that Jews had 'too much influence' (Miller *et al.*, op. cit.). Jirinovsky's father was a Jew.

13 O. Böss, *Die Lehre der Eurasier. Ein Beitrag zur russischen Ideengeschichte des 20. Jahrhunderts*, Wiesbaden, Otto Harrassowitz, 1961.

14 A. Vichnevski, 'Le nationalisme russe: à la recherche du totalitarisme perdu', *Hérodote*, 1994, vol. 72/73, pp. 101–18; D. Kerr, 'The new Eurasianism'.

15 H. Adomeit, 'Russia as a "great power" in world affairs: images and reality', *International Affairs*, 1995, vol. 71, pp. 50ff.

16 R. Brubaker, 'Nationhood and the national question in the Soviet Union and post-Soviet Eurasia: an institutionalist account', *Theory and Society*, 1995, vol. 23, pp. 47–78.

17 Brubaker, 'Nationhood and the national question', p. 62.

18 Adomeit, 'Russia as a "great power"'.

19 V. Chernov, 'Significance of the Russian military doctrine', *Comparative Strategy*, 1994, vol. 13, pp. 161–6.

20 Kerr, 'The new Eurasianism', p. 978.

21 Z. Brzezinski, 'The premature partnership', *Foreign Affairs*, 1994, vol. 73, p. 80

22 A.Z. Rubinstein, 'The geopolitical pull on Russia', *Orbis*, 1994, vol. 38, pp. 567–84.

23 Hauner, *What Is Asia to us?*, p. 107.
24 S. Blank, 'Russia's real drive to the south', *Orbis*, 1995, vol. 39, pp. 369–86.
25 V. Kolossov, A. Treivish and R. Tourovsky, 'Les systèmes géopolitiques d'Europe orientale et centrale vus de la Russie', in E. Philippart (ed.), *Nations et frontières dans la nouvelle Europe*, Brussels, Éditions Complexes, 1993.
26 The proportions of territory distributed between Europe and Asia have remained close to those of the Soviet Union: 1:3. Kerr, 'The new Eurasianism', p. 981.
27 Borodaj and Nikiforov, 'Between East and West', pp. 112–13. Trotsky had already voiced the same ideas.

CHAPTER 9: THE EMPIRE OF REVENGE
(Serbia)

1 This text was previously published (in Dutch) in *Internationale Spectator*, 1993, vol. 47, pp. 488–92. The current version has some minor additions and an Afterword.
2 M.E. Durham, *Twenty Years of Balkan Tangle*, London, Allen & Unwin, 1920.
3 M.E. Durham, *Through the Lands of the Serb*, London, Edward Arnold, 1904, p. 302.
4 I. Banac, 'The fearful asymmetry of war: the causes and consequences of Yugoslavia's demise', *Daedalus*, 1992, vol. 121, no. 1, p. 142.
5 M. Ignatieff, 'The Balkan tragedy', *New York Review of Books*, 13 May 1993.
6 M. Glenny, *The Fall of Yugoslavia: The Third Balkan War*, London, Penguin Books, 1992, p. 41.
7 C. Jelavich and B. Jelavich, *The Establishment of the Balkan National States, 1804–1920*, Seattle, University of Washington Press, 1977, p. 261.
8 J. Vidmar, *Zwischen Verzicht und Behauptung*, Klagenfurt, Drava Verlag, 1984.
9 P.R. Lawrence and J.W. Lorsch, *Organization and Environment*, Homewood, Harvard University Press, 1969.
10 Serbia became a European laughing-stock because of the quarrels between King Milan and his wife Natalija in the 1880s and the assassination of King Alexander, his wife Draga – allegedly a prostitute – and members of Draga's family in 1903. According to Durham's report, the military officers responsible for the assassination pitched the bodies (still breathing) from the window and defiled them where they had fallen, even cutting portions of Draga's skin which they dried and preserved as trophies. Durham, *Twenty Years*.
11 Jelavich and Jelavich, *The Establishment*, p. 259.
12 G. Stokes, *Politics as Development: The Emergence of Political Parties in Nineteenth-Century Serbia*, Durham/London, Duke University Press, 1990, p. 293.
13 More precisely his analysis of the paralysis of the Action Party. See A. Gramsci, 'Notes on Italian history', in A. Gramsci, *Selections from the Prison Notebooks*, New York, International Publishers, 1971, pp. 44–120.
14 Durham, *Twenty Years*, p. 239.
15 R. Marković, 'What are Yugoslavia's internal borders?' in *The Creation and Changes of the Internal Borders of Yugoslavia*, Ministry of Information of the Republic of Serbia, 1991(?).
16 W. Vermeer, 'Over grenzen, genocide en historische belangen: een brief van de Servische oppositieleider aan de Kroatische president', *Internationale Spectator*, 1991, vol. 45, pp. 406–9.
17 J. Ilić, 'Characteristics and importance of some ethno-national and political-geographic factors relevant for the possible political-legal disintegration of Yugoslavia', in *The Creation and Changes of the Internal Borders of Yugoslavia*, Ministry of Information of the Repulic of Serbia, 1991.
18 Written in January 1996.
19 M. van de Port, *Het Einde van de Wereld*, Amsterdam, Babylon-de Geus, 1994.
20 The text between quotation marks is mine.

CHAPTER 10: TOTALLY LOST?
(Iraq)

1 This text was previously published (in Dutch) in *Internationale Spectator*, 1994, vol. 48, pp. 63–7.

2 There is no common opinion on the precise year. Different sources give AD 635, 636, 637 or 638. F.M. Donner, *The Early Islamic Conquests*, Princeton, Princeton University Press, 1981, p. 176.

3 The earliest referral, to my knowledge, is in a speech of Saddam Hussein eight days after the start of the war (28 September 1980). See J.M. Abdulghani, *Iraq and Iran: The Years of Crisis*, Baltimore, Johns, Hopkins University Press, 1984, p. 208.

4 J.L. Richardson, *The Gulf War and Australian Political Culture*, Canberra, Australian National University, 1992, p. 19.

5 K. Makiya, *Cruelty and Silence: War, Tyranny and Uprising in the Arab World*, London, Jonathan Cape, 1993, p. 159.

6 See, for a carefree rendition of the myth, J.B. Glubb, *The Great Arab Conquests*, London, Hodder & Stoughton, 1963 (used as a basis for my own rendition in this chapter). This is not an authorized report of historical facts but in the context of myths all variants are interesting. See, for a cautious report of the historical facts: F.M. Donner, *The Early Islamic Conquests*. A more extensive treatment of this battle is in S.M. Yusuf, 'The battle of al-Qadisiyya', *Islamic Culture*, 1945, vol. 19, pp. 1–28.

7 The same obtains for the Iraqi image of the Hindus: see Samir al-Khalil (Kanan Makiya), *Republic of Fear*, New York, Pantheon Books, 1989, pp. 120–1. See also the reaction of the Palestine/Jordanian journalist Khouri on the participation of Egypt in the anti-Iraq alliance in the war to liberate Kuwait: 'Egypt, as always, is an exception because of its obsequious political servitude.' R.G. Khouri, 'The bitter fruits of war', in M.L. Sifry and C. Cerf, *The Gulf War Reader*, New York, Times Books, 1991, p. 403.

8 S. Bromley, *Rethinking Middle East Politics: State Formation and Development*, Cambridge, Polity Press, 1994, p. 174.

9 A. Baram, *Culture, History and Ideology in the Formation of Ba'thist Iraq 1968–89*, New York, St Martin's Press, 1991.

10 See, for an overview of political symbols in postage stamps, D.M. Reid, 'The postage stamp: a window on Saddam Hussayn's Iraq', *Middle East Journal*, 1993, vol. 47, pp. 77–89.

11 Baram, *Culture, History and Ideology*, pp. 42–3.

12 V.S. Naipaul, *Among the Believers: An Islamic Journey*, London, Penguin Books, 1982, p. 350. This particular remark concerned an Indonesian Muslim.

13 Baram, *Culture, History and Ideology*, p. 27.

14 A. Baram, 'Mesopotamian identity in Ba'thi Iraq', *Middle Eastern Studies*, 1983, vol. 19, pp. 426–55.

15 Abdulghani, *Iraq and Iran*, p. 78.

16 Makiya, *Cruelty and Silence*, pp. 255–7.

17 W. Khalidi, 'Iraq vs Kuwait: claims and counterclaims', in Sifry and Serf, *The Gulf War Reader*, p. 57.

18 T.Y. Ismael and J.S. Ismael, 'Arab politics and the Gulf war: political opinion and political culture', *Arab Studies Quarterly*, 1993, vol. 15, pp. 1–11. See also Makiya, *Cruelty and Silence*.

19 Ismael and Ismael 'Arab politics'.

20 R. Hartshorne, 'The functional approach in political geography', *Annals of the Association of American Geographers*, 1950, vol. 40, pp. 95–130.

21 Samir al-Khalil, *Republic of Fear*, p. 120.

22 An indication that political leaders may also fall victim to this uncertainty can be found in the political memoirs of Ba'th politician Sami al-Jundi who, shortly before his flight to Tunis, played a prominent part in Syrian politics: 'In the whirlpool of

March [1963] you thought yourself to be exercising responsibility, planning and making history, but you became suddenly nothing. A little while later you saw yourself again as very important, and then, again, nothing.' E. Kedourie, *Arabic Political Memoirs*, London, Frank Cass, 1974, p. 203.

23 Samir al-Khalil, *Republic of Fear*, p. 116.
24 A. Richards and J. Waterbury, *A Political Economy of the Middle East: State, Class and Economic Development*, Boulder, Westview Press, 1990, p. 184.
25 Naipaul, *Among the Believers*, p. 112.

CHAPTER 11: A WORLD IN ITSELF
(India)

1 J. Nehru, *The Discovery of India*, New Delhi, Oxford University Press, 1981 (1st edn 1946), pp. 539–40.
2 Nehru, *The Discovery of India*, p. 533.
3 A. Varshney, 'Contested meanings: India's national identity, Hindu nationalism, and the politics of anxiety', *Daedalus*, 1993, vol. 122 no. 3, pp. 234.
4 Unless one considers the traditional definition of the 'Hindu' land, a term referring in Sanskrit to both river (Indus) and sea (Shindu), as that territory lying between the Himalayas, the Indus and the ocean, to be something so obvious that it does not require comment.
5 Nehru, *The Discovery of India*, p. 62.
6 A. Embree, *Utopias in Conflict: Religion and Nationalism in Modern India*, Berkeley, University of California Press, 1990, p. 79.
7 V.S. Naipaul, *India: A Million Mutinies Now*, London, Minerva, 1990, p. 8.
8 H. Kapur, *India's Foreign Policy 1947–1992: Shadows and Substance*, New Delhi, Sage, 1994, pp. 120–2.
9 Kapur, *India's Foreign Policy*, p. 181.
10 V.K.V.R. Rao, *The Nehru Legacy*, 1971. Quoted in Kapur, *India's Foreign Policy*, pp. 120–1.
11 Nehru, *The Discovery of India*, p. 56.
12 As he did in the US and the Soviet Union. Nehru, *The Discovery of India*, p. 56.
13 S. Wolpert, *An Introduction to India*, New Delhi, Penguin Books, 1990, p. 241.
14 D.P. Moynihan, *Pandaemonium: Ethnicity in International Politics*, Oxford, Oxford University Press, 1993, p. 157.
15 A small nuclear test had already taken place in 1974.
16 Wolpert, *An Introduction to India*, p. 19.
17 A.B. Shah, 'Public opinion in Indian democracy', 1965, quoted in Kapur, *India's Foreign Policy*, p. 167.
18 Kapur, *India's Foreign Policy*, p. 168.
19 P.S. Gosh, 'Foreign policy and electoral politics in India: inconsequential connection', *Asian Survey*, 1994, vol. 34, pp. 807–17.
20 Gosh, 'Foreign policy', p. 813.
21 Gosh, 'Foreign policy', p. 815.
22 J.-L. Racine, 'Identité nationale et *Realpolitik*: entre Chine et Occident, l'Inde de 1989 face au printemps de Pékin', *Hérodote*, 1993, vol. 71, pp. 217–40.
23 Racine, 'Identité nationale', p. 224.
24 A. Nandy, 'The political culture of the Indian state', *Daedalus*, 1989, vol. 118, no. 4, pp. 23–4.
25 Nandy, 'The political culture', p. 24.
26 G. Bocqerat, 'L'Inde dans l'après-guerre froide', *Hérodote*, 1993, vol. 71, pp. 203–16.
27 J. Bakshi, 'India's geopolitical importance in the United Nations', *India Quarterly: A Journal of International Affairs*, 1994, vol. 50, p. 95.

CHAPTER 12: CONCLUSION

1 'Security risk' may be conceived as synonymous with 'geopolitical vision', which would make the statement tautological. By security risk, I simply mean a real, inevitable change in the international pattern of interaction.

2 H.-H. Nolte, 'The alleged influence of cultural boundaries on political thinking: images of Central Europe', in A. Gerrits and N. Adler, *Vampires Unstaked: National Images, Stereotypes and Myths in East Central Europe*, Amsterdam, North-Holland, 1995, pp. 41–54.

3 A Dutch newspaper (*NRC-Handelsblad*) scored 12 per cent in 1984. R. Verheij, *De Invloed van Buitenlandse Politiek op Buitenlands Nieuws. Een Vergelijking van het Amerikabeeld in Nederland en Zweden* (The Influence of Foreign Policy on Foreign News: A Comparison of the Image of America in the Netherlands and Sweden), unpublished MA thesis, University of Amsterdam, 1988.

4 I. Budge and D.J. Farlie, *Explaining and Predicting Elections: Issue Effects and Party Strategies in Twenty-Three Democracies*, London, Allen & Unwin, 1983. Figures concern elections between 1946 and 1983.

5 D. Corbey, *Een Krimpende Horizon of een Wijder Perspektief* (A Shrinking Horizon or a Wider Perspective), unpublished MA thesis, University of Amsterdam, 1987. The countries involved were Britain, Spain and the Netherlands.

6 C. Harvie, *The Rise of Regional Europe*, London, Routledge, 1994.

7 H. van der Wusten, 'Les Symboles de la future carte géopolitique de l'Europe', in E. Philippart (ed.), *Nations et frontières dans la nouvelle Europe*, Brussels, Éditions Complexe, 1993, pp. 127–38.

BIBLIOGRAPHY

Abdulghani, J.M., *Iraq and Iran: The Years of Crisis*, Baltimore, Johns Hopkins University Press, 1984.

Adomeit, H., 'Russia as a "great power" in world affairs: images and reality', *International Affairs*, 1995, vol. 71, pp. 35–86.

Alger, C.F., 'Perceiving, analysing and coping with the local–global nexus', *International Social Science Journal*, 1988, vol. 117, pp. 321–39.

—— 'The world relations of cities: closing the gap between social science paradigm and everyday human experience', *International Studies Quarterly*, 1990, vol. 34, pp. 493–518.

—— 'Local response to global intrusions', in Z. Mlinar (ed.), *Globalization and Territorial Identities*, Aldershot, Avebury, 1992, pp. 77–104.

Allison, G. and Treverton, G.F., *Rethinking America's Security: Beyond Cold War to New World Order*, New York, W.W. Norton, 1992.

Anderson, P., 'Nation-states and national identity', *London Review of Books*, 9 May 1991.

Anonymous (G.T. Chesney), *The Battle of Dorking: Reminiscences of a Volunteer*, Edinburgh, William Blackwood and Sons, 1871.

Appleyard, B., 'The dinkum Aussie? Strewth?', *Independent*, 8 September 1993.

Ashbaugh, C., 'Judge Gary vs. the people: conspiracy trials in America', in D. Roediger and F. Rosemont (eds), *Haymarket Scrapbook*, Chicago, C.H. Kerr, 1986, pp. 243–46.

Atherton, J., 'The vocabulary of the Vietnam war', in J.-R. Rougé, *L'Opinion américaine devant la guerre du Vietnam*, 1992.

Avrich, P., *The Haymarket Tragedy*, Princeton, Princeton University Press, 1984.

—— 'The bomb-thrower: a new candidate', in D. Roediger and F. Rosemont (eds), *Haymarket Scrapbook*, Chicago, C.H. Kerr, 1986, pp. 71–3.

Bakker, G., *Duitse Geopolitiek 1919–1945: een Imperialistische Ideologie*, Assen, Van Gorcum, 1967.

Bakshi, J., 'India's geopolitical importance in the United Nations', *India Quarterly: A Journal of International Affairs*, 1994, vol. 50, pp. 93–8.

Banac, I., 'The fearful asymmetry of war: the causes and consequences of Yugoslavia's demise', *Daedalus: Journal of the American Academy of Arts and Sciences*, 1992, vol. 121, no. 1, 141–74.

Baram, A., 'Mesopotamian identity in Ba'thi Iraq', *Middle Eastern Studies*, 1983, vol. 19, pp. 426–55.

—— *Culture, History and Ideology in the Formation of Ba'thist Iraq 1968–89*, New York, St Martin's Press, 1991.

Bartlett, C.J., *British Foreign Policy in the Twentieth Century*, Basingstoke, Macmillan, 1989.

Bassin, M., 'Russian geographers and the "national mission" in the Far East', in D. Hooson (ed.), *Geography and National Identity*, Oxford, Blackwell, 1994, pp. 112–33.

Bean, C., 'Are Australian attitudes to government different?: A comparison with five other nations', in F.G. Castles, *Australia Compared*, 1991, pp. 74–100.

Bell, P.M.H., *John Bull and the Bear: British Public Opinion, Foreign Policy and the Soviet Union 1941–1945*, London, Edward Arnold, 1990.

Blackwell, M., *Clinging to Grandeur: British Attitudes and Foreign Policy in the Aftermath of the Second World War*, Westport, Greenwood Press, 1993.

Blank, S., 'Russia's real drive to the south', *Orbis*, 1995, vol. 39, pp. 369–86.

Bloom, W., *Personal Identity, National Identity and International Relations*, Cambridge, Cambridge University Press, 1990.

Blum, D.W., 'The Soviet foreign policy belief system: beliefs, politics and foreign policy outcomes', *International Studies Quarterly*, 1993, vol. 37, pp. 373–94.

Böss, O., *Die Lehre der Eurasier. Ein Beitrag zur russischen Ideengeschichte des 20. Jahrhunderts*, Wiesbaden, Otto Harrassowitz, 1961.

Boquerat, G., 'L'Inde dans l'après-guerre froide', *Hérodote*, 1993, vol. 71, pp. 203–16.

Borodaj, J. and Nikiforov, A., 'Between East and West: Russian renewal and the future', *Studies in East European Thought*, 1995, vol. 47, pp. 61–116.

Boulding, K., *The Image*, Ann Arbor, University of Michigan Press, 1956.

Boyle, P.G., 'The British foreign office view of Soviet–American relations, 1945–46', *Diplomatic History*, 1979, vol. 3, pp. 307–20.

Braudel, F., *The Identity of France*, vol. II, *People and Production*, London, Collins, 1990.

Breuilly, J. (ed.), *The State of Germany: The National Idea in the Making, Unmaking and Remaking of a Modern Nation-State*, London, Longman, 1992.

Bromley, S., *Rethinking Middle East Politics: State Formation and Development*, Cambridge, Polity Press, 1994.

Brubaker, R., 'Nationhood and the national question in the Soviet Union and post-Soviet Eurasia: an institutionalist account', *Theory and Society*, 1995, vol. 23, pp. 47–78.

Brunner, R.D., 'Myth and American politics', *Policy Sciences*, 1994, vol. 27, pp. 1–18.

Brzezinski, Z., 'The premature partnership', *Foreign Affairs*, 1994, vol. 73, pp. 67–82.

Budge, I. and Farlie, D.J., *Explaining and Predicting Elections: Issue Effects and Party Strategies in Twenty-Three Democracies*, London, Allen & Unwin, 1983.

Burleigh, M., 'Scholarship, state and nation 1918–1945', in J. Breuilly (ed.), *The State of Germany*, London, Longman, 1992.

Butor, M., 'L'espace du roman', in M. Butor, *Répertoire II*, Paris, Éditions de Minuit, 1964.

Calleo, D.P., *The German Problem Reconsidered: Germany and the World Order, 1870 to the Present*, Cambridge, Cambridge University Press, 1978.

Campbell, D., *Writing Security: United States Foreign Policy and the Politics of Identity*, Manchester, Manchester University Press, 1992.

Carasales, J., *National Security Concepts of States: Argentina*, New York, UNIDIR/United Nations, 1992.

Carter, P., *Living in a New Country: History, Travelling and Language*, London, Faber & Faber, 1992.

Castles, F.G., *Australia Compared: People, Policies and Politics*, Sydney, Allen & Unwin, 1991.

Chernov, V., 'Significance of the Russian military doctrine', *Comparative Strategy*, 1994, vol. 13, pp. 161–6.

Child, J., *Geopolitics and Conflict in South America*, New York, Praeger, 1985.

Chomsky, N., 'Visions of righteousness', in J.C. Rowe and R. Berg (eds), *The Vietnam War and American Culture*, 1991.

Christmas, L., *The Ribbon and the Ragged Square: An Australian Journey*, London, Penguin Books, 1987.

Clarke, I.F., *Voices Prophesying War: Future Wars 1763–3749*, Oxford, Oxford University Press, 1992.

Cohen, S.B., *Geography and Politics in a World Divided*, New York, Random House, 1963.

Colley, L., *Britons: Forging the Nation 1707–1837*, Yale, Yale University Press, 1992.

Collins, H., 'Political ideology in Australia: the distinctiveness of a Benthamite society', *Daedalus: Journal of the American Academy of Arts and Sciences*, 1985, vol 114, no. 1, pp. 147–69.

Colmer, J., *Australian Autobiography: The Personal Quest*, Melbourne, Oxford University Press, 1989.

Cooper, A.F., 'Between fragmentation and integration: the evolving security discourse in Australia and Canada', *Australian Journal of International Affairs*, 1995, vol. 49, pp. 49–67.

Corbey, D., *Een Krimpende Horizon of een Wijder Perspektief*, Amsterdam, unpublished MA thesis, University of Amsterdam, 1987.

Corradi, J.E., 'Menem's Argentina, Act II', *Current History*, 1995, vol. 94, pp. 76–80.

Crassweller, R., *Perón and the Enigmas of Argentina*, New York, W.W. Norton, 1987.

Cutts Dougherty, K., Eisenhart, M. and Webley, P., 'The role of social representations and national identities in the development of territorial knowledge: a study of political socialization in Argentina and England', *American Educational Research Journal*, 1992, vol. 29, pp. 809–36.

Dalby, S., 'American security discourse: the persistence of geopolitics', *Political Geography Quarterly*, 1990, vol. 9, pp. 171–88.

Dodds, K.-J., 'Creating a strategic crisis out of a communist drama? Argentine and South African geo-graphs of the South Atlantic', *European Review of Latin American and Caribbean Studies*, 1994, vol. 56, pp. 33–54.

—— 'Geopolitics and foreign policy: recent developments in Anglo-American political geography and international relations', *Progress in Human Geography*, 1994, vol. 18, pp. 186-208.

Donner, F.M., *The Early Islamic Conquests*, Princeton, Princeton University Press, 1981.

Draper, T., 'McNamara's peace', *New York Review of Books*, 11 May 1995.

Duncan, T. and Fogarty, J., *Australia and Argentina: On Parallel Paths*, Melbourne, Melbourne University Press, 1984.

Durham, M.E., *Twenty Years of Balkan Tangle*, London, Allen & Unwin, 1920.

Eisner, M., 'Long-term dynamics of political values in international perspective: comparing the results of content analysis of political documents in the USA, GB, FRG and Switzerland', *European Journal of Political Research*, 1990, vol. 18, pp. 605–21.

Eksteins, M., *Rites of Spring: The Great War and the Birth of the Modern Age*, New York, Doubleday, 1989.

Embree, A., *Utopias in Conflict: Religion and Nationalism in Modern India*, Berkeley, University of California Press, 1990.

Ferraro, D., *La Personalidad cultural argentina*, Buenos Aires, Editora de la Palmera, 1985.

Filipp, K., *Germany Sublime and German Sublimations: On Political Education and its Geography*, Munster, Waxmann, 1993.

Fitzgerald, F., *Fire in the Lake: The Vietnamese and the Americans in Vietnam*, Boston, Little Brown and Company, 1972.

Foster, D.W., *The Argentine Generation of 1880: Ideology and Cultural Texts*, Columbia, University of Missouri Press, 1990.

Fukuyama, F., *The End of History and the Last Man*, London, Hamish Hamilton, 1992.

Gaddis, J.L., *Strategies of Containment: A Critical Appraisal of Postwar American National Security Policy*, New York, Oxford University Press, 1982.

Garreau, J.L., *The Nine Nations of North America*, New York, Avon Books, 1982.

Garton Ash, T., *In Europe's Name: Germany and the Divided Continent*, New York, Random House, 1993.

Gerchunoff, A., *El Hombre importante*, Buenos Aires, Colección et Pasado Argentino, 1960.

Gerrits, A. and Adler, N., *Vampires Unstaked: National Images, Stereotypes and Myths in East Central Europe*, Amsterdam, North-Holland, 1995.

Gleason, J.H., *The Genesis of Russophobia in Great Britain: A Study of the Interaction of Policy and Opinion*, Cambridge, Harvard University Press, 1950.

Glenny, M., *The Fall of Yugoslavia: The Third Balkan War*, London, Penguin Books, 1992.

Glubb, J.B., *The Great Arab Conquests*, London, Hodder & Stoughton, 1963.

Gorbachev, M., *Perestroika: New Thinking for our Country and the World*, London, Collins, 1987.

Gosh, P.S., 'Foreign policy and electoral politics in India: inconsequential connection', *Asian Survey: A Monthly Review of Contemporary Asian Studies*, 1994, vol. 34, pp. 807–17.

Gourley, J., Saarinen, T.F. and MacCabe, C., 'Comparison of sketch maps drawn by students of Armidale, Australia and Dunedin, New Zealand', *New Zealand Journal of Geography*, 1993, vol. 96, pp. 8–15.

Gramsci, A., *Selections from the Prison Notebooks* (Q. Hoare and G. Nowell Smith, eds), New York, International Publishers, 1971.

Green, D., *The Language of Politics in America: Shaping Political Consciousness from McKinley to Reagan*, Ithaca, Cornell University Press, 1992.

Gruner, W.D., *Die deutsche Frage: Ein Problem der europäischen Geschichte seit 1800*, Münch, C.H. Beck, 1985.

Gurry, M., 'Identifying Australia's region: from Evatt to Evans', *Australian Journal of International Affairs*, 1995, vol. 49, pp. 17–31.

Hahn, K.-E., 'Westbindung und Interessenlage. Über die Renaissance der Geopolitik', in H. Schwilk and U. Schacht (eds), *Die selbstbewusste Nation*, pp. 327–44.

Harper, J.L., *American Visions of Europe: Franklin D. Roosevelt, George F. Kennan and Dean G. Acheson*, Cambridge, Cambridge University Press, 1994.

Hartshorne, R., 'The functional approach in political geography', *Annals of the Association of American Geographers*, 1950, vol. 40, pp. 95–130.

Harvie, C., *The Rise of Regional Europe*, London, Routledge, 1994.

Haselbach, D., 'Political correctness: zur gegenwärtigen politischen Kultur in Nordamerika', *Neue Politische Literatur*, 1995, vol. 40, pp. 116–33.

Hauner, M., *What Is Asia to Us? Russia's Asian Heartland Yesterday and Today*, London, Routledge, 1992.

Hawke, B., *The Hawke Memoirs*, London, Heinemann, 1994.

Hayden, R.M., 'The use of national stereotypes in the wars in Yugoslavia', in A. Gerrits and N. Adler, *Vampires Unstaked*, 1995, pp. 207–22.

Hellmann, J., *American Myth and the Legacy of Vietnam*, New York, Columbia University Press, 1986.

Hepple, L.W., 'The revival of geopolitics', *Political Geography Quarterly*, 1986, vol. 5 (suppl.), pp. 21–36.

Herman, E.S. and Chomsky, N., *Manufacturing Consent: The Political Economy of the Mass Media*, New York, Vintage, 1988.

Higgott, R., 'The politics of Australia's international economic relations: adjustment and two level games', *Australian Journal of Political Science*, 1991, vol. 26, pp. 2–28.

Hiro, D., *The Longest War*, London, Paladin Grafton Books, 1989.

Holsti, O.R., 'The belief system and national images: a case study', *Journal of Conflict Resolution*, 1962, vol. 6, pp. 244–52.

Holsti, O.R. and Rosenau, J.N., 'The structure of foreign policy beliefs among American opinion leaders – after the cold war', *Millennium*, 1993, vol. 22, pp. 235–78.

Hooson, D. (ed.), *Geography and National Identity*, Oxford, Blackwell, 1994.

Hopf, T., *Peripheral Visions: Deterrence Theory and American Foreign Policy in the Third World 1965–1990*, Ann Arbor, University of Michigan Press, 1994.

Hughes, R., *Culture of Complaint*, New York, Warner Books, 1994.

Ignatieff, M., 'The Balkan tragedy', *New York Review of Books*, 13 May 1993.

Ilić, J., 'Characteristics and importance of some ethno-national and political-geographic factors relevant for the possible political-legal disintegration of Yugoslavia', in *The Creation and Changes of the Internal Borders of Yugoslavia*, Ministry of Information of the Republic of Serbia, 1991(?).

Inglehart, R., *Culture Shift in Advanced Industrial Society*, Princeton, Princeton University Press, 1990.

Ismael, T.Y. and Ismael, J.S., 'Arab politics and the Gulf war: political opinion and political culture', *Arab Studies Quarterly*, 1993, vol. 15, pp. 1–11.

James, H., *A German Identity: 1770 to the Present Day*, London, Phoenix, 1989/1994.

Jelavich, C. and Jelavich, B., *The Establishment of the Balkan National States, 1804–1920*, Seattle, University of Washington Press, 1977.

Jervis, R., *Perception and Misperception in International Politics*, Princeton, Princeton University Press, 1976.

Johnson, P., 'The myth of American isolationism', *Foreign Affairs*, 1995, vol. 74, pp. 159–64.

—— 'Blood on the wattle', *Times Literary Supplement*, 13 May 1994, p. 25.

Jose, N. 'Cultural identity: "I think I'm something else"', *Daedalus: Journal of the American Academy of Arts and Sciences*, 1985, vol. 114, no. 1, pp. 311–42.

Kappeler, A., 'Some remarks on Russian national identities (sixteenth to nineteenth centuries)', *Ethnic Groups*, 1993, vol. 10, pp. 147–55.

Kapur, H., *India's Foreign Policy 1947–1992: Shadows and Substance*, New Delhi, Sage, 1994.

Kasperson, R.E. and Minghi, J., *The Structure of Political Geography*, London, University of London Press, 1969.

Kates, R.W., 'Experiencing the environment as hazard', in S. Wapner, S.B. Cohen and B. Kaplan, *Experiencing the Environment*, New York, Plenum Press, 1976.

Kedourie, E., *Arabic Political Memoirs*, London, Frank Cass, 1974.

Kelly, P. and Child, J., *Geopolitics of the Southern Cone and Antarctica*, Boulder, Lynne Riener, 1988.

Kerr, D., 'The new Eurasianism: the rise of geopolitics in Russia's foreign policy', *Europe-Asia Studies*, 1995, vol. 47, pp. 977–88.

Khalidi, W., 'Iraq vs Kuwait: claims and counterclaims', in M.L. Sifry and C. Cerf, *The Gulf War Reader*, 1991.

Khalil, Samir al-, (pseudonym of Kanan Makiya), *Republic of Fear*, New York, Pantheon Books, 1989.

Khouri, R.G., 'The bitter fruits of war', in M.L. Sifry and C. Cerf, *The Gulf War Reader*, 1991.

King, J., *Waltzing Materialism – and Other Attitudes that have Shaped Australia 1788–1978*, Sydney, Harper & Row, 1978.

Kissinger, H.A., *The White House Years*, London, Weidenfeld & Nicolson/Michael Joseph, 1979.

—— (ed.) *Report of the National Bipartisan Commission on Central America*, Washington, US Government Printing Office, 1984.

—— 'Balance of power sustained', in G. Allison and G.F. Treverton (eds), *Rethinking America's Security*, New York, W.W. Norton, 1992, pp. 238–48.

—— 'We live in an age of transition', *Daedalus: Journal of the American Academy of Arts and Sciences*, 1995, vol. 124, no. 3, pp. 99–107.

Klingberg, F.L., 'Cyclical trends in American foreign policy moods and their policy implications', in C.W. Kegley Jr and P.J. McGowan (eds), *Challenges to America*, Beverly Hills, Sage, 1979, pp. 37–55.

Kolossov, V., Treivish, A. and Tourovsky, R., 'Les systèmes géopolitiques d'Europe orientale et centrale vus de la Russie', in E. Philippart (ed.), *Nations et frontières dans la nouvelle Europe*, Brussels, Éditions Complexes, 1993.

Kovel, J., *Red Hunting in the Promised Land: Anticommunism and the Making of America*, New York, Basic Books, 1994.

Kundera, M., 'The tragedy of Central Europe', *New York Review of Books*, 26 April 1984.

Lacoste, Y., *La Géographie ça sert d'abord à faire la guerre*, Paris, Maspéro, 1976.

Lanús, J.A., *La Causa argentina*, Buenos Aires, Emecé Editores, 1988.

Lawrence, P.R. and Lorsch, J.W., *Organization and Environment: Managing Differentiation and Integration*, Homewood, Harvard University Press, 1969.

Leaver, R., 'Biting the dust: the imperial conventions within republican pretences', *Australian Journal of Political Science*, 1993, vol. 28, pp. 146–61.

Leland, C.T., *The Last Happy Men: The Generation of 1922, Fiction and the Argentine Reality*, Syracuse, Syracuse University Press, 1986.

Loprete, C.A., *El Ensueño argentino*, Buenos Aires, Editorial Plus Ultra, 1985.

Lowenthal, D., 'Geography, experience and imagination: towards a geographical epistemology', *Annals of the Assciation of American Geographers*, 1961, vol. 51, pp. 241–60.

—— 'The American scene', *Geographical Review*, 1968, vol. 58, pp. 61–88.

Mackinder, H.J., 'The geographical pivot of history', *The Geographical Journal*, 1904, vol. 23, pp. 421–44.

McKinley, M., 'To be "kept at all hazards": the British construction, and spatio-temporal deconstruction of Ireland as subdued threat and strategic asset', *History of European Ideas*, 1994, vol. 19, pp. 101–13.

Mafud, J., *Psicología de la viveza criolla: Contribuciones para una interpretación de la realidad social argentina y americana*, Buenos Aires, Editorial Americalee, 1965.

Makiya, K., *Cruelty and Silence: War, Tyranny and Uprising in the Arab World*, London, Jonathan Cape, 1993.

Mann, K., *Distinguished Visitors. Der Amerikanische Traum*, Munich, Edition Spangenberg, 1992 (1939).

Marković, R., 'What are Yugoslavia's internal borders?' in *The Creation and Changes of the Internal Borders of Yugoslavia*, Ministry of Information of the Republic of Serbia. 1991(?).

Marquand, D., 'After Whig imperialism: can there be a new British identity?', *New Community*, 1995, vol. 21, pp. 183–93.

Mead, W.R., 'The once and future Reich', *World Policy Journal*, 1990, vol. 7, pp. 593–638.

Meany, N., 'The end of "White Australia" and Australia's changing perception of Asia, 1945–1990', *Australian Journal of International Affairs*, 1995, vol. 49, pp. 171–89.

Mechtersheimer, A., 'Nation und Internationalismus. Über nationales Selbst-bewusstsein als Bedingung des Friedens', in H. Schwilk and U. Schacht (eds), *Die selbstbewusste Nation*, pp. 345–63.

Melanson, R.A., *Reconstructing Consensus: American Foreign Policy since the Vietnam War*, New York, St Martin's Press, 1991.

Mertes, M., 'Germany's social and political culture: change through consensus?', *Daedalus: Journal of the American Academy of Arts and Sciences*, 1994, vol. 123, no. 1, pp. 1–32.

Millar, J.B., *Australia in Peace and War*, Botany, Australian National University Press, 1991, 2nd edn.

Miller, A.H., Hesli, V.L. and Reisinger, W.M., 'Comparing citizen and elite belief systems in post-Soviet Russia and Ukraine', *Public Opinion Quarterly*, 1995, vol. 59, pp. 1–40.

Miller, D., 'Reflections on British national identity', *New Community*, 1995, vol. 21, pp. 153–66.

Mlinar, Z. (ed.), *Globalization and Territorial Identities*, Aldershot, Avebury, 1992.

Moynihan, D.P., *Pandaemonium: Ethnicity in International Politics*, Oxford, Oxford University Press, 1993.

Mueller, J.E., *War, Presidents and Public Opinion*, New York, University Press of America, 1973.

Naipaul, V.S., *The Return of Eva Peron: With the Killings in Trinidad*, London, André Deutsch, 1980.

—— *Among the Believers: An Islamic Journey*, London, Penguin Books, 1982.

—— *India: A Million Mutinies Now*, London, Minerva, 1990.

Nandy, A., 'The political culture of the Indian state', *Daedalus: Journal of the American Academy of Arts and Sciences*, 1989, vol. 118, no. 4, pp. 1–26.

Newman, J. and Unsworth, M., *Future War Novels: An Annotated Bibliography of Works in English Published since 1946*, Phoenix, Oryx Press, 1984.

Nijinski, R., *Nijinski*, London, Victor Gollancz, 1933.

Nijman, J. and Wusten, H. van der, 'Breaking the cold war mould in Europe: a geopolitical tale of gradual change and sharp snaps', in J. O'Loughlin and H. van der Wusten (eds), *The New Political Geography of Eastern Europe*, 1993, pp. 15–30.

Nolte, H.-H., 'The alleged influence of cultural boundaries on political thinking: images of Central Europe', in A. Gerrits and N. Adler, *Vampires Unstaked*, 1995, pp. 41–54.

Okhrimenko, V., 'Jirinovski et "le bond final de la Russie sur le Sud"', *Hérodote*, 1994, vol. 72/73, pp. 119–27.

O'Loughlin, J. and Grant, R., 'The political geography of presidential speeches 1946–1987', *Annals of the Association of American Geographers*, 1990, vol. 80, pp. 504–30.

O'Loughlin, J. and Wusten, H. van der(eds), *The New Political Geography of Eastern Europe*, London, Belhaven Press, 1993.

Ossenbrügge, J., 'Territorial ideologies in West Germany 1945–1985: between geopolitics and regionalist attitudes', *Political Geography Quarterly*, 1989, vol. 8, pp. 387–99.

Papadakis, E., 'Does the new politics have a future?', in F.G. Castles, *Australia Compared*, 1991, pp. 239–57.

Platt, D.C.M. and di Tella, G., *Argentina, Australia and Canada: Studies in Comparative Development 1870–1965*, Oxford, Macmillan, 1986.

Port, M. van de, *Het einde van de wereld. Beschaving, redeloosheid en zigeunercafé's in Servië*, Amsterdam, Babylon-de Geus, 1994.

Porter, B., *Britain, Europe and the World 1850–1986: Delusions of Grandeur*, London, Allen & Unwin, 1986.

Racine, J.-L., 'Identité nationale et *Realpolitik*: entre Chine et Occident, l'Inde de 1989 face au printemps de Pékin', *Hérodote*, 1993, vol. 71, pp. 217–40.

Raeff, M., 'The people, the intelligentsia and Russian political culture', *Political Studies*, 1993, vol. 41, pp. 93–106.

Ratzel, F., *Deutschland. Einführung in die Heimatkunde*, Leipzig, Fr. Wilhelm Grunow, 1898.

Reid, D.M., 'The postage stamp: a window on Saddam Hussayn's Iraq', *Middle East Journal*, 1993, vol. 47, pp. 77–89.

Renouf, A., *The Frightened Country*, Melbourne, Macmillan, 1979.

Richards, A. and Waterbury, J., *A Political Economy of the Middle East: State, Class and Economic Development*, Boulder, Westview Press, 1990.

Richardson, J.L., *The Gulf War and Australian Political Culture*, Canberra, Research School of Pacific Studies, Australian National University, 1992.

Robertson, J.O., *American Myth, American Reality*, New York, Hill & Wang, 1980.

Rock, D., *Authoritarian Argentina: The Nationalist Movement, its History and its Impact*, Berkeley, University of California Press, 1993.

Roediger, D. and Rosemont, F. (eds), *Haymarket Scrapbook*, Chicago, C.H. Kerr, 1986.

Röhl, K.R., 'Morgenthau und Antifa. Über den Selbsthass der Deutschen', in H. Schwilk and U. Schacht (eds), *Die selbstbewusste Nation*, 1994, pp. 96–7.

Rooney Jr, C.J., *Dreams and Visions: A Study of American Utopias, 1865–1917*, Westport, Greenwood Press, 1985.

Rössler, M., *'Wissenschaft und Lebensraum'. Geographische Ostforschung im Nationalsozialismus*, Berlin/Hamburg, Dietrich Reimer Verlag, 1990.

Rougé, J.-R. (ed.), *L'Opinion américaine devant la guerre du Vietnam*, Paris, Presses de l'Université de Paris-Sorbonne, 1991.

Rowe, J.C. and Berg, R. (eds), *The Vietnam War and American Culture*, New York, Columbia University Press, 1991.

Rubinstein, A.Z., 'The geopolitical pull on Russia', *Orbis*, 1994, vol. 38, pp. 567–84.

Rudolph, J.D., *Argentina: A Country Study*, Washington DC, Headquarters Department of the Army, 1986.

Said, E.W., *Orientalism*, London, Routledge & Kegan Paul, 1978.
—— 'Afterword to the 1995 printing', in E.W. Said, *Orientalism*, London, Penguin Books, 1995.
Sanders, D. and Edwards, G., 'Consensus and diversity in elite opinion: the views of the British foreign policy elite in the early 1990s', *Political Studies*, 1994, vol. 42, pp. 413–40.
Saunders, D., 'What makes a nation a nation? Ukrainians since 1600', *Ethnic Studies*, 1993, vol. 10, pp. 101–24.
Schlesinger, A.M. Jr, *The Cycles of American History*, Boston, Houghton Mifflin, 1986.
—— *The Disuniting of America*, New York, W.W. Norton, 1992.
Schlesinger, A. Jr, 'Back to the Womb? Isolationism's renewed threat', *Foreign Affairs*, 1995, vol. 74, pp. 2–8.
Schmidt, H., *Menschen und Mächte*, Berlin, Siedler Verlag, 1987.
Schoultz, L., *National Security and United States Policy toward Latin America*, Princeton, Princeton University Press, 1987.
Schulte-Sasse, J., 'Karl Mays Amerika-Exotik und deutsche Wirklichkeit. Zur sozialpsychologischen Funktion von Trivialliteratur im wilhelminischen Deutschland', in H. Schmiedt, *Karl May*, Frankfurt am Main, Suhrkamp, 1983, pp. 101–29.
Schultz, H.-D., 'Fantasies of *Mitte*: *Mittellage* and *Mitteleuropa* in German geographical discussion in the 19th and 20th centuries', *Political Geography Quarterly*, 1989, vol. 8, pp. 315–39.
Schulz, E., *Die deutsche Nation in Europa. Internationale und historische Dimensionen*, Bonn, Europa Union Verlag, 1982.
Schwilk, H. and Schacht, U. (eds), *Die selbstbewusste Nation. 'Anschwellender Bocksgesang' und weitere Beiträge zu einer deutschen Debatte*, Berlin, Ullstein, 1994.
Scobie, J.R., *Buenos Aires: Plaza to Suburb, 1870–1910*, New York, Oxford University Press, 1974.
Shapiro, M.J., *The Politics of Representation: Writing Practices in Biography, Photography, and Policy Analysis*, Madison, University of Wisconsin Press, 1988.
Sharp, J.P., 'Publishing American identity: popular geopolitics, myths and The Readers Digest', *Political Geography*, 1993, vol. 12, pp. 491–503.
Shumway, N., *The Invention of Argentina*, Berkeley, University of California Press, 1991.
Sifry, M.L., and Cerf, C., *The Gulf War Reader*, New York, Times Books, 1991.
Sloan, G.R., *Geopolitics in United States Strategic Policy, 1890–1987*, Brighton, Wheatsheaf, 1988.
Slotkin, R., *Regeneration through Violence: The Mythology of the American Frontier, 1600–1860*, Middletown, Wesleyan University Press, 1973.
—— *The Fatal Environment: The Myth of the Frontier in the Age of Industrialization 1800–1980*, New York, Atheneum, 1985.
Smith, A.D., *National Identity*, London, Penguin Books, 1991.
—— 'National identity and the idea of European unity', *International Affairs*, 1992, vol. 68, pp. 55–76.
Smith, S.K. and Wertman, D.A., *US–West European Relations during the Reagan Years: The Perspective of West European Publics*, London, Macmillan, 1992.
Spangenberg, D., *Die blockierte Vergangenheit: Nachdenken über Mitteleuropa*, Berlin, Argon, 1987.
Spiering, M., *Englishness: Foreigners and Images of National Identity in Postwar Literature*, PhD thesis, University of Amsterdam, 1993.
Sprout, H. and Sprout, M., 'Environmental factors in the study of international politics', *Journal of Conflict Resolution*, 1957, vol. 1, pp. 309–28.
Stokes, G., *Politics as Development: The Emergence of Political Parties in Nineteenth-Century Serbia*, Durham/London, Duke University Press, 1990.
Strassoldo, R., 'Globalism and localism: theoretical reflections and some evidence', in Z. Mlinar (ed.), *Globalization and Territorial Identities*, 1992.
Swales, M., 'The problem of nineteenth-century German realism', in N. Boyle and M.

Swales, *Realism in European Literature: Essays in Honour of J.P. Stern*, Cambridge, Cambridge University Press, 1986.

Tamamoto, N., 'The ideology of nothingness: a meditation on Japanese national identity', *World Policy Journal*, 1994, vol. 11, pp. 89–99.

Taylor, A.J.P., *The Course of German History: A Survey of the Development of Germany since 1815*, London, Hamish Hamilton, 1945; Methuen, 1978.

Taylor, P.J., *Britain and the Cold War: 1945 as Geopolitical Transition*, London, Pinter, 1990.

Thies, J., 'Perspektiven deutscher Aussenpolitik', in R. Zitelmann, K. Weissmann and M. Grossheim (eds), *Westbindung*, 1993, pp. 523–36.

Tilly, C., *Coercion, Capital and European States: AD 990–1992*, Cambridge, Blackwell, 1990.

Tuathail, G. Ó, 'The language and nature of the "new geopolitics" – the case of US–El Salvador relations', *Political Geography Quarterly*, 1986, vol. 5, pp. 73–85.

Tuathail, G. Ó and Agnew, J., 'Geopolitics and discourse: practical geopolitical reasoning in American foreign policy', *Political Geography*, 1992, vol. 11, pp. 190–204.

Tucker, R.W. and Hendrickson, D.C., *The Imperial Temptation: The New World Order and America's Purpose*, New York, Council on Foreign Relations Press, 1992.

Urban, M., 'The politics of identity in Russia's postcommunist transition: the nation against itself', *Slavic Review*, 1994, vol. 53, pp. 733–65.

Varenne, H., 'The question of European nationalism', in T.M. Wilson and M. Estellie Smith (eds), *Cultural Change and the New Europe*, Boulder, Westview Press 1993.

Varlin, T., 'La Mort de Che Guevara: les problèmes du choix d'un théâtre d'opérations en Bolivie', *Hérodote*, 1977, vol. 5, pp. 39–81.

Varshney, A., 'Contested meanings: India's national identity, Hindu nationalism, and the politics of anxiety', *Daedalus: Journal of the American Academy of Arts and Sciences*, 1993, vol. 122, no. 3, pp. 227–62.

Verheij, R., *De Invloed van Buitenlandse Politiek op Buitenlands Nieuws. Een Vergelijking van het Amerikabeeld in Nederland en Zweden*, MA thesis, University of Amsterdam, 1988.

Vermeer, W., 'Over grenzen, genocide en historische belangen: een brief van de Servische oppositieleider aan de Kroatische president', *Internationale Spectator*, 1991, vol. 45, pp. 406–9.

Vichnevski, A., 'Le nationalisme russe: à la recherche du totalitarisme perdu', *Hérodote*, 1994, vol. 72/73, pp. 101–18.

Vidmar, J., *Zwischen Verzicht und Behauptung. Essays zur Identitätsfindung des slowenischen Volkes*, Klagenfurt, Drava Verlag, 1984.

Wagner, C., *Die Tagebücher (1)* M. Gregor-Dellin and D. Mack, (eds), Munich/Zurich, R. Piper, 1976.

Waisman, C.H., *Reversal of Development in Argentina: Postwar Counterrevolutionary Policies and their Structural Consequences*, Princeton, Princeton University Press, 1987.

Wallace, W., 'Foreign policy and national identity in the United Kingdom', *International Affairs*, 1991, vol. 67, pp. 65–80.

—— 'British foreign policy after the cold war', *International Affairs*, 1992, vol. 68, pp. 423–42.

Wanders, A.C., Nicolas, G. and Parker, G., *Géovision allemande des Europes*, Lausanne, Eratosthène, 1995.

Warhurst, J., 'Nationalism and republicanism in Australia: the evolution of institutions, citizenship and symbols', *Australian Journal of Political Science*, 1993, vol. 28, pp. 100–20.

Welsh, J., 'The role of the inner enemy in European self-definition: identity, culture and international relations theory', *History of European Ideas*, 1994, vol. 19, pp. 53–61.

Wilensky, H., *Organizational Intelligence*, New York, Basic Books, 1967.

Wills, G., 'The new revolutionaries', *New York Review of Books*, 10 August 1995.

Wilson, T.M. and Estellie Smith, M. (eds), *Cultural Change and the New Europe*, Boulder, Westview Press, 1993.

Wippermann, W., *Der 'Deutsche Drang nach Osten'. Ideologie und Wirklichkeit eines politischen Schlagwortes*, Darmstadt, Wissenschaftliche Buchgesellschaft, 1981.

Wittkopf, E.R., *Faces of Internationalism: Public Opinion and American Foreign Policy*, Durham, Duke University Press, 1990.

Wollschläger, H., *Karl May in Selbstzeugnisse und Bilddokumente*, Reinbek/Hamburg, Rowohlt, 1965.

Wolpert, S., *An Introduction to India*, New Delhi, Penguin Books, 1990.

Wusten, H. van der, 'Les Symboles de la future carte géopolitique de l'Europe', in E. Philippart (ed.), *Nations et frontières dans la nouvelle Europe*, Brussels, Éditions Complexe, 1993, pp. 127–38.

Wynia, G., *Argentina: Illusions and Realities*, New York, Holmes & Meier, 1986.

Xiang, L., 'Is Germany in the West or in Central Europe?' *Orbis*, 1992, vol. 36, pp. 411–42.

Yusuf, S.M., 'The battle of al-Qadisiyya', *Islamic Culture*, 1945, vol. 19, pp. 1–28.

Zitelmann, R., Weissmann, K. and Grossheim, M. (eds), *Westbindung. Chancen und Risiken für Deutschland*, Frankfurt and Main, Propyläen Verlag, 1993.

INDEX

Note: page references in **bold** show the major entry

Abalkin, Leonid 98
Abkhazia 106
Acheson, Dean 61
'adaptation level' 3
Adomeit, Hannes 103
Afghanistan invaded by Britain 40
Afghanistan invaded by Soviet Union
    15, 31, 92, 143; and India 135;
    retreat from 134; and Russia/Soviet
    Union 104, 105; and USA 63, 70
Africa 84–5, 101, 137
Aidid, General 71
Akkadians 121
Albania 109, 112, 116
Alberdi, Juan Bautista 75–6
Alger, Chadwick F. 7
Algiers, treaty of 123
Amalrik, Andrej 101
Amerindians see Indians, South
    American
Amis, Kingsley 37, 38
Amritsar, Golden Temple of 131
Amur, River 8, 100
anarchism see terrorism and anarchism
Anderson, Perry 38
Anglo-French-Russian Entente 41
Anglo-Japanese treaty 41
Anglo-Russian Entente 40, 41
Anglo-Saxons see Australia; Britain;
    United States
Angst 35; see also Germany
Antarctica 82, 85
anti-materialism 21–2
anti-semitism 21
ANZAC pact 91
ANZUS treaty 91, 92
appeasement 62
Appleyard, Brian 87
Arabia 119, 120

Arabs see Gulf War; Iraq; Islam; Middle
    East; Orientalism
archaeology 121
'archetypes' 1
Arctic 98
Arent, Alfred 73
Argentina **72–85**, 115, 156–7; and
    borders 85; and Britain 73, 75, 78,
    80–5; and Central America 85; and
    Christianity 79, 83–4, 85; economy
    72–5, 77–8, 80–1; and Europe 72–3,
    75–6, 78–84; freedom, idea of 75–8;
    geopolitical visions 139, 141, 144;
    God as Argentinian 72–5; and
    imperialism, Spanish 75, 76, 79, 81,
    82, 83; and Italy 73, 79, 157; and
    Japan 72, 73; metamorphosis of state
    78–81; models, using other countries
    as 12; national myths and
    international perceptions 83–5;
    national self-analysis 81–3; and
    Russia/Soviet Union 79, 80; test of
    boys' territorial and national
    knowledge 2; and United States 72,
    73, 80, 83, 85; and World Wars 80,
    84
armed forces: Argentina 73, 79; India
    132, 133; Russia/Soviet Union 105,
    106; see also wars
Armenia 117
Art and culture 87; high 25, 33; music
    20–1, 24, 25
Ash, T. Garton 30–1, 32
Asia: and Australia 86–7, 90, 91; British
    colonies in 39, 128, 129, 132, 133,
    137; geopolitical visions 139, 140,
    141, 143, 144, 145, 146; industry
    15; and Russia/Soviet Union 15, 98,
    101, 105, 106–8; see also China;

India; Japan; Orientalism; Pakistan; Russia/Soviet Union; South-east Asia and Pacific
attitude surveys *see* public opinion surveys
Aubin, Hermann 27
Australia 42, **86–94**, 158–9; and Asia 86–7, 90, 91; and Britain 86, 87, 88, 89, 90–1, 93; and China 90–1, 92; and Cold War 91, 93; economy 74, 75, 86, 93–4; and Europe 89, 93–4; geography versus history 86–7; and geopolitical visions 90–3, 139, 140, 141, 144, 145, 146; and Gulf War 90, 91; and identity 88–90, 94; and Japan 89, 101; and myths and symbolism 88–90, 93; national character, problem of 87–8; and new realism 93–4; and USA 88, 89, 90–4; and World Wars 89, 91, 92
Austria 19, 29, 88
Austria-Hungary and Habsburgs 28, 115, 116
authoritarianism 87
autocracy *see* dictatorship and autocratic rule
Ayarragaray (Argentinian writer) 74, 77
Ayodhya Mosque 130
Azerbaijan 106

Babri Mosque 130
Babylon 121
Baden-Württemberg 10
Bail, Murray 90
balance of power 60–1, 62, 101; *see also* Cold War; Eurasia concept; Heartland
Balkans 106, 110, 116; Albania 109, 112, 116; Bulgaria 106, 114, 116; Greece 60, 106, 112, 142; *see also* Bosnia; Croatia; Serbia
Baltic area 98, 101, 143
Banac, Ivo 110, 112, 113, 114
Bangladesh 133
Baram, A. 122, 124
Bartlett, C.J. 39
Basil III, Tsar of Russia 100
Ba'th Party 121
Bayreuth festival 21, 25
Beagle Channel 85
Bedouin 119
Beitritt 30
Belarus 101, 106, 108
Belgium 10, 28
Belgrade 110, 112
belief constraint 96–7

belief-system theory 14–15; *see also* geopolitical visions
Bell, P.M.H. 43
Bellamy, Edward 53
Belarussians 106
Bengal 133
Bentham, Jeremy 93
Berlin 23, 24; Wall *see* 'Wende'
Bevin, Ernest 41, 42, 46
Bharatiya Janata Party (India) 134, 136, 137
Bildung 22
Bismarck, Otto Edward Leopold von 4, 19, 23, 24, 28
BJP *see* Bharatiya Janata Party
Black Hand Society 109, 115
Blackwell, Michael 46
Bloom, William 7
Blum, Douglas 14, 15
Bolivia 12, 14, 31
Bolshevik Revolution 96, 100, 102, 140
bombing: Oklahoma City explosion 49, 50–1, 56–7; spatial pattern in Vietnam War 4; *see also* terrorism and anarchism
Borodaj, Jurij 99–100, 107
borders: and Argentina 85; and France 19; and geopolitical visions 144; and Germany 19, 26–9, 34; and Italy 19; in language of foreign policy 5; in mental maps of world 1–2; as 'natural' 12; natural, lack of 19; Russia/Soviet Union ('common houses') 97–8, 101; and Serbia 115; and Spain 19; US concepts of 1–2, 49, 51–3, 55–7, 67
Borges, Jorge Luis 81
Bosch, Robert 10
Bosnia 102, 112, 116, 117, 115
'boundary-producing' phenomenon: foreign policy as 5, 51; *see also* borders
Brahmins 134
Brandt, Willy 30, 32
Braudel, F. 27, 38
Brazil 72, 82, 83, 85
Britain **36–48**, 152–3; and Argentina 73, 75, 78, 80–5; and Asia 39, 41, 128, 129, 132, 133, 137; and Australia 86, 87, 88, 89, 90–1, 93; Battle of Dorking 36–8, 41, 152; and China 12, 47; and Cold War 41–4; economy 24, 46–7; and EU 38, 46–7, 93; fiction 36–8, 41, 44–5, 152; and France 37, 39, 41; geopolitical heritage of 19th century 38–41; geopolitical transition of 20th century

41–4; geopolitical visions 11–12, 141, 142, 144–5; and Germany 36, 37, 39; ideology of sovereignty 44–8; and India 128, 129, 132, 133, 137; and Iran 40; and Japan 41, 47; literature 23; as model 12, 20; nuclear bases, protest against 7; and Russia/Soviet Union 39–40, 41, 43, 95, 105, 143; and Serbia 110; and Suez crisis 44; test of boys' territorial and national knowledge 2; and Turkey 39, 40, 42; and USA 28, 38, 42–4, 46, 84, 140; as villain in German fiction 22; and World Wars 42, 45; *see also* imperialism of Britain
Brubaker, Rogers 103
Brzezinski, Zbigniew 105
Buenos Aires 72, 75, 79, 81, 83
Bulgaria 106, 114, 116
Bush, George 58, 71, 90
Butor, Michel 6, 10

Campbell, D. 5, 54, 55
Canada 94, 98
Carter, Jimmy 63
casualties of wars 61–2, 71, 89, 111, 123, 126
Catalonia 145
Catholics *see* Roman Catholicism
Caucasus 105, 106
caudillismo 75, 156
Central America 131; and Argentina 85; and USA 5, 31, 63–5, 71
Central Asia republics 105–6
Central Europe *see* Eastern Europe
centrality of Germany *see* Mitteleuropa concept
change and continuity in geopolitical visions 139–41
Charter '77 32
Chechnya 106, 107
Chernobyl 32, 146
Chesney, George T. 36, 37
Chetniks 112, 116
Chicago 24; *see also* Haymarket riot
Child, Jack 83
childhood 2, 16
Chile 82, 83, 85
China: and Australia 90–1, 92; and Britain 12, 47; communism 32; Cultural Revolution 7–8; empire foreseen 24; as enemy in fiction 70; Great Leap Forward 7–8; and Hong Kong 12; and India 131–3, 134, 136–7, 138, 140; isolationism 63; and Pakistan 137; and Russia/Soviet

Union 4, 15, 63, 101, 106, 137; and USA 60, 90–1; and Vietnamese 65
Chomsky, Noam 31, 64
'chosen people' 99
Christaller, Walter 27–8
Christianity 8–9, 59; and Argentina 79, 83–4, 85; fundamentalist 58; and Russia/Soviet Union 96, 98–101, 102, 105, 106; and Serbia 112, 114
Christmas, Linda 87
Churchill, Sir Winston 41, 42, 43
cinema *see* fiction
CIS *see* Commonwealth of Independent States
city: Argentina 72, 75, 79, 81, 83; as core area 12, 52; as frontier town 49; Germany 23, 24; and rest of world 6, 10; and Russia/Soviet Union 100, 101, 105, 106; USA 10, 24, 49
'Civic Union' coalition 97
Civil Wars: America 51, 53, 57, 61; Greece 60; Pakistan 133; Spain 79; theme of US future wars fiction 68, 69–70; Yugoslavia 3 (*see also* Serbia)
Clark, Manning 87, 158
Clarke, I.F. 70
class-consciousness 87
classical geopolitics 3–4
Clinton, Bill 58, 71
code, geopolitical 12
'coexistence', system-opening 29–32
Cold War 32; and Australia 91, 93; and Britain 41–4; and geopolitical visions 139–44 *passim*; and India 136; nuclear deterrence 3–4; and Russia/Soviet Union 98, 99, 101; and USA 5, 16, 54–6, 57, 62; end of *see* 'Wende'
collective mission (national mission) 12, 63; *see also* geopolitical visions
Colley, Linda 37
colonialism *see* imperialism
'common houses' 97–8, 101
'commonality' of culture, economy, law and duty 11
Commonwealth, British 46; *see also* Australia; Britain *under* imperialism
Commonwealth of Independent States 34, 104, 106, 107
communication gap between Vietnam and USA 65–6
communism: Bolshevik Revolution 96, 100, 102, 140; expansion *see* domino theory; South Atlantic possibilities 85; uprisings against 9, 14; demise of 16, 57, 71, 138 (*see also* 'Wende'); *see also*

China; Eastern Europe; fear of communism; Marx; Russia/Soviet Union
Congress Party (India) 130, 137
conspiracy fantasies in Iraq 123
Constantinople 100, 105
constitutional patriotism 34
consumerism in India 138
containment policy of USA 60–1
continuity and change in geopolitical visions 139–41
Coppola, Francis F. 67
core area 12, 29, 52; see also Heartland
core-beliefs see belief-system
Corsica 107
Ćosić, Dobrica 111, 112
countries, national identity of see in particular Argentina; Australia; Britain; Germany; Russia/Soviet Union; Serbia; United States
Crèvecoeur 51
Crimean War 40
critical geopolitics 4
Croatia 110, 111, 112, 114, 115, 116, 117
Cuba 12–13, 64
Cultural Revolution (China) 7–8
culture: art see Art and Culture; gaps between nations
see also Orientalism; rest of worldmass 11
currency 34
cycles in history 58
Czechoslovakia 32; invasion of 9, 14, 135, 143

dangers, external 5, 9; and Argentina 79–80, 84; British invasion fears 36, 37, 39; invoked from third nation by rapport of two 16; terrorism and anarchism 49, 50, 51; and USA 49, 50, 51; see also borders; fear; immigrants; wars
Danube, River (and basin) 28–9
Darwin, Charles/Darwinism 20, 25
Dayton Accord 117
democracy 84; lack of 127; Russia/Soviet Union 95, 96; see also party politics
dependency 104; economic independence 80; myth in Australia 93
Depression, Great 73–4, 75, 78, 80
Desert Shield, Operation 91
Desert Storm, Operation 63, 120
détente and Nixon 63

deterrence see nuclear weapons
Diaghilev, Sergei 72
dictatorship and autocratic rule: in Argentina 74, 75–6, 78, 84; in Europe see Hitler; in Russia 96, 100 (see also Stalin)
Dimitrijević, Colonel 115
discourse and experience 2–3
diversity and unity in India 128–31
Dnieper, River 105
domestic rationality in Iraq 126–7
domino theory of communist expansion 14, 42, 61, 133, 155
Don, River 105
Donnelly (British consul) 43, 44
Dorking, Battle of 36–8, 41, 152
Dostoevskij, Fyodor 100
Drakulić, Slavenka 111
Draper, Theodore 155
Dravidian movement 130–1
Dreyfus, Alfred 39
drug trade as external danger 5
Duarte, Eva (Péron) 81
Dublin 10
Dubrovnik 112
Durham, M. Edith 110, 115, 116, 160
duties, common 11
dying for country, people less willing 10; for higher cause 89

East Germany 17, 30; see also 'Wende'
East and West, difference between 7, 98; self-definition as bridge between (Germany, Russia and India) 141, 142; see also Cold War; Orientalism
Eastern Europe 17; and geopolitical visions 142, 143; and Germany 27, 28, 30–2, 34, 35; and India 125; intelligentsia cut off 9; and Russia/Soviet Union 14, 101, 106, 135, 143; see also Czechoslovakia; Hungary; Poland; Serbia; Yugoslavia
economy and trade 23–4; Argentina 72–5, 77–8, 80–1; Australia 74, 75, 86, 93–4; Britain 24, 46–7; common 11; cooperation see European Union; free enterprise 14; free trade 46–7; and geopolitical visions 145, 147; Russia/Soviet Union 95, 104; and state 145
education, lack of 135
Egypt 39, 44, 105; and Iraq 120, 126, 127, 161
Eisenhower, Dwight D. 61
El Salvador 31
Elchibey (Azerbaijan President) 106

elections: to European Parliament 10;
　see also party politics and elections
elite: India 137–8; see also middle class
Embree, A. 130
emigration see immigrants
émigré circles 103
Empire see imperialism
Engels, Friedrich 24
Enlightenment 75, 83, 97, 100
environmental problems 32, 146
'equilibrium' concept 62–3
Ethiopia 120
'ethnic cleansing' see genocide
EU see European Union
Eurasia concept: core 29; dilemma see
　Russia/Soviet Union; as enemy in US
　fiction 70; and Russia/Soviet Union
　15, 102, 106–8, 140, 144, 145; and
　world balance of power 60–1, 62; see
　also Heartland
'Eurocentrism' 10
Europe (mainly Western): and Argentina
　72–3, 75–6, 78–84; and Australia
　89, 93–4; emigration from see
　immigrants; and India 138; industry
　and quality 15; levels of identity and
　interpretation 6–11; low level of
　national pride 10; materialism 23–6,
　28, 88; migration from see
　immigrants; migration to 9; novels on
　future wars 69; public opinion
　fluctuations 11; and Russia/Soviet
　Union 96, 97–8, 99, 100, 101, 104,
　105, 107; and Serbia 116; and USA
　50–7, 59; and Yugoslavian civil war
　3; see also Britain; Eastern Europe;
　European Union; France; Germany;
　Netherlands
European Parliament 10
European Union/European Community:
　and Britain 38, 46–7, 93; and
　geopolitical visions 139, 145; and
　Germany 33, 34; low level of national
　pride 10; migration 9; protectionism
　94; Single European Act 107; see also
　Europe (mainly Western)
evangelism of fear see witch-hunt
Evil Empire concept 63
experience and discourse 2–3
external factors see dangers; invasion;
　rest of world

Falklands see Malvinas
family and morality ideals in USA 57, 58
fascism 79, 112; see also National
　Socialism

'Fashoda Incident' 39
fear: of communism see Cold War and
　Vietnam War and also witch-hunt; of
　nuclear reactor accidents 32, 146; of
　nuclear weapons 144; pathological
　regime of 126; see also dangers,
　external; invasion; security; terrorism
　and anarchism
Federal Bureau of Investigation 57
Federal Republic of Germany see
　Germany
Ferraro, Diana 81–3
Festspiele see Bayreuth festival
fiction (mainly novels and films) 23,
　144–5; Argentina 76, 81; Britain
　36–8, 41, 44–5, 152; future wars 37,
　67–70, 156; Germany 20–2, 24, 76;
　lack of 157; modern novel 6; see also
　literature; myths and symbolism
First World War: and Argentina 84; and
　Australia 89; and Britain 45;
　casualties 61, 89, 111; and
　geopolitical visions 141; and Germany
　18, 22, 23, 26, 29; and India 132;
　and Russia/Soviet Union 101; and
　Serbia 110, 111, 116; and USA 53–4,
　61; end of see Versailles
Fitzgerald, Frances 53, 65–6
Foreign Office (Britain) 38, 39, 42, 143
foreign policy and affairs: as
　'boundary-producing' phenomenon 5,
　51; and geopolitical visions 11; and
　idealism 60; language of 5;
　prominence of 142–4; US beliefs
　analysed 64–5; see also belief-system;
　geopolitical visions; rest of world
'Fortress Europe' 9
Foucault, Michel 7
France: and Argentina 73, 80, 84; and
　Berlin Wall 1; borders 19; and Britain
　37, 39, 41; and geopolitical visions
　146; and Germany 33; imperialism 8,
　28; industry 24; literature 23; low
　level of national pride 10; materialism
　88; as model 12, 20; modern novel 6;
　Napoleonic 1, 15, 37, 38, 86, 99;
　political geography 4, 5; Revolution
　29, 84; and Russia/Soviet Union 41;
　and Suez crisis 44; as villain in
　German fiction 22; and Wagner's
　music 21
Franz Ferdinand, Archduke of Austria
　116
Fraser, John M. 92
free enterprise 14
'Free hand' principle 38

free trade 46–7
freedom, Argentinian idea of 75–8
FRG (Federal Republic of Germany) *see* Germany
Friuli 10
frontiers *see* borders
Fukuyama, F. 95
fundamentalism: Christian 58; Islamic 14, 49, 57, 106; Sikh 131
future: balance of power 101; of Russia/Soviet Union, ambiguous 95–9; wars, fiction about 37, 67–70, 156

Gaddis, John L. 12
Gallipoli 89
Gandhi, Indira 131, 134, 135
Gandhi, Mahatma 132
Gandhi, Rajiv 131, 134, 136, 137
gaucho myth 76, 89
GDR (German Democratic Republic) *see* East Germany
Gemeinschaft 23
genius of place 103
genocide and 'ethnic cleansing' 3, 154; in former Yugoslavia 109, 111, 112, 113–14, 116, 117–18; U.S. war practice 61
geography: 'imaginative' 3; politics of *see* geopolitical; geopolitics; 'sacred' 129, 130; versus history in Australia 86–7; zonation 102; *see also* national experience of place; rivers
geopolitical code 12
geopolitical reflex 99; absent in Russia/Soviet Union 103–6
geopolitical visions 9, 10–15, **139–47**; Australia 90–3; continuity and change 139–41; defined 11, 16; foreign affairs, prominence of 142–4; missions and national identity 146–7; nature of 141–2; security 144–6; and World Wars 18, 22, 23, 26, 29; *see also* rest of world
geopolitics 1; academic decline into disrepute 4; classical and modern 3–6; defined 3; language of 5; *see also* geopolitical visions; political geography
Geopolitik 1
Georgia 106
Gerchunoff, Alberto 77
German Democratic Republic *see* East Germany
German-Russian non-aggression treaty 29

Germany **17–35**, 150–2; borders 19, 26–9, 34; and Britain 36, 37, 39; central location and *Mitteleuropa* concept 14, 17–20, 24, 28–9, 33, 35, 144, 151; division of 17 (*see also* '*Wende*'); and Eastern Europe 27, 28, 30–2, 34, 35; emigration to USA 50; and EU 33, 34; fiction 20–2, 24, 76; and France 33; geopolitical visions 140–6 *passim*; history 25–6, 28; imperialism 24, 27–8; low level of national pride 10; materialism and idealism 23–6, 28, 88; models, using other countries as 12; occupation of 9, 17, 46; Orientalism 8; and Poland 26–7, 31, 32; post-war views of 16; reaction and rancour 20–3; reunification *see* '*Wende*'; and Russia/Soviet Union 17, 28, 29–32, 72, 73, 79, 80, 82, 98, 101, 104; system-opening 'coexistence' 29–32; and USA 17, 31, 32, 33, 34; and World Wars 18, 22–3, 26, 28, 29, 30, 34, 35; after '*die Wende*' 32–5; *see also* Hitler; National Socialism
Gesamtkunstwerk 25
Giliaks 8
Gladstone, William 38
Glaspie, April 70, 126
Gleason, J.H. 40
Glenny, Misha 110
God: as impersonal force 14; special dispensation to Argentina 72–5; *see also* religion
Gorbachev, Mikhail 14, 97–8, 100, 137, 154
Gosh, Partha S. 136
Gramsci, A. 115
Grand Slam, Operation 133
Great Leap Forward: China 7–8
Great Patriotic War (Russia/Soviet Union) 98, 102
Great Russia concept 102
Greater Serbia 112–17
Greece 106, 112, 142; Civil War 60
Green Berets 52–3
Greenham Common protest 7
'groupthink' 123
Gründerzeit 20
guerilla warfare 12–14, 52
Guevara, Che 12–14, 53
guilt, German collective feeling of 33
Gulf War 17, 120, 126; and Australia 90, 91; casualties 123, 126; and India 135–6; Iraqi rationale for 123,

125; and Russia/Soviet Union 102;
and USA 63, 70–1, 73
Gurtal (Indian Foreign Minister) 136

Habsburgs *see* Austria-Hungary
Hamad, Mahmud 122, 124
Harriman, Averell 42–3
Hartshorne, Richard 126
Harvie, Christopher 145
Hauner, Milan 15, 105
Haushofer, Karl 3, 27, 29, 35
Hawke, Bob 90, 91, 92
Haymarket (Chicago) riot 49–50, 51,
52, 53, 56
Heartland and Rimland concept: and
Argentina 72, 73; and Britain 40; and
Germany 29; and India 128; and USA
49, 56, 60, 63; *see also* core area;
Mackinder
hegemony as power politics 64
Hellmann, John 67
Herman, E.S. 31
Hernández, José 76
Hervé, Jean 13
hidden meaning of geographical data 4
Hinduism 129–30, 134–6, 162;
Hinduization 136
history: common 11; cycles 58;
geopolitic, symbolism of 25–6; and
Germany 25–6, 28; and Iraq 120–3;
and Serbia 109–11, 112; versus
geography in Australia 86–7
Hitler, Adolf 71; aims and strategy 29,
42, 46; appeased 62; fascism of *see*
National Socialism; and geopolitical
theory 3; and May's fiction 21;
Russian foreign policy and 98, 99
Holland *see* Netherlands
holy places 12, 99, 120
Holy Roman Empire 28
'Holy Russia' mission 100–1
Honecker, Erich 31
Hong Kong 12, 47
hostage crisis in Iran 60, 63
House, Edward 54
Hughes, Robert 57–8
human rights and violations: and
Australia 80; in China *see*
Tiananmen; in Iraq 8, 62; and USA
31, 63
Hungary 114; Rising of 1956 and Soviet
invasion 9, 14, 135, 143; *see also*
Austria-Hungary
Hussein, Saddam: as benefactor 124;
and human rights violation 62; and
India 136; and Nebuchadnezzar

120–3; and terror system 144; and
USA 70, 71, 120, 125; and war with
Iran 119, 120–3, 141; *see also* Gulf
War

idealism: and foreign policy 60, 62; and
materialism in Germany 23–6; and
USA 2, 57, 88; *see also* myth
identity (mainly national identity): in
Australia 88–90, 94; fear of losing 38,
44–5; German search for 22; Indian
130–1, 133–6; levels and
interpretation 6–11; loss with change
of location 66; and negativity 22, 89;
and religion in South Asia 130–1,
132, 133–6; Russia/Soviet Union
99–102; USA 51–3, 140; violation
57–8; *see also* local experience and
identification; national identity
ideology: Australian avoidance of 87;
Ideological and Imperial paradigm
103, 104; India 130–1, 132, 133–6;
of sovereignty (Britain) 44–8; *see also*
beliefs
Ignatieff, Michael 110
'imaginative geography' 3
IMF (International Monetary Fund) 136
immigrants: to Argentina from Europe
74, 76, 79–80, 83; to Europe 9; to
USA from Europe 50, 51, 59
imperialism 8–9, 141; in Argentina 75,
76, 79, 81, 82, 83; of Britain 8–9, 24,
28, 37, 39–40, 42, 45–6, 47,
128–30, 132, 133, 137; of Germany
24, 27–8; in India 128–30, 132; and
Ottoman Empire 39, 40, 100; and
Russia/Soviet Union 99–100, 103,
104; of Spain *see* Argentina *above*
impersonal forces, assumption of 14
India 39, **128–38**, 162; and Britain
128, 129, 132, 133, 137; and China
131–3, 134, 136–7, 138, 140; and
Cold War 136; and Eastern Europe
125; and Europe 138; and geopolitical
visions 138, 139, 141, 142, 144,
145, 146; and Gulf War 135–6; and
identity 130–1, 133–6; and ideology
130–1, 132, 133–6; and Islam
129–30, 133, 136; and Pakistan 132,
133; and partition of 129–30 (*see also*
Pakistan); and public opinion 135–8,
143; and rest of world 131–4; and
Russia/Soviet Union 40, 98, 105,
128, 133, 134, 136; and unity and
diversity 128–31; and USA 128, 134,
136, 137; and wars 128, 132, 133

Indian Ocean 98, 101, 105, 136
Indians, South American 75, 76, 82
individualism 49, 83
Indochina War 132; *see also* Vietnam
  War
industrial growth *see* economy and trade
inevitability 24
Innerlichkeit 23
institutions, political, Germany's lack of
  22
intermediary, self-image as 141, 142
intermediate beliefs 14
'intermestic' issues 136
International Geographical Congress 27
International Monetary Fund (IMF)
  136
'Internet' culture 3
intervention 63–4; *see also* Vietnam War
invasion 101; British fears 36, 37, 39,
  144–5; Eastern Europe 9, 14, 135,
  143; Kuwait *see* Gulf War; *see also*
  Afghanistan
Inverchapel, Lord 43
invincibility, Serbian feeling of 116,
  Muslim armies 133
Iran (Persia) 62, 119, 123; and Britain
  40; geopolitical visions 144; and
  Russia/Soviet Union 42, 44, 106; US
  hostage crisis 60, 63; war with Iraq
  120, 122, 126, 141
Iraq 62, **119–27**, 161–2; domestic
  rationality 126–7; and Egypt 120,
  126, 127, 161; geopolitical visions
  141, 144; historical heritage 120–3;
  human rights violations in 8; and
  India 136; and Islam 119–20, 121,
  127; myth and symbolism 119–20,
  121, 122; Qadisiyya, battles of
  119–20, 121; regional rationality
  123–6; and Turkey 127; war with
  Iran 120, 122, 126, 141; *see also* Gulf
  War
Ireland 10, 44, 153
iron curtain concept 42
Islam: fundamentalism 14, 49, 57, 106;
  and India 129–30, 133, 136;
  internationalism 121; and Iraq
  119–20, 121, 127; Mecca 12, 120;
  and Russia/Soviet Union 99, 102,
  106; and Serbia 109, 112, 114–15,
  116; Shiites and Sunnis 121, 126; *see
  also* Gulf War; Middle East; Ottoman
  Empire
Ismael, T.Y. and J.S. 161
isolationism and isolation 14; Argentina
  80; Australia 92, 93; Britain 39;

Chinese 63; Germany 18; India 132,
  135–7; USA 51–2, 53, 54, 59–60; *see
  also* rest of world
Israel 102; and Australia 89; and Iraq
  121, 123; and Wagner's music 21
Italy 115; and Argentina 73, 79, 157;
  borders 19; local attachment 10;
  materialism 88; and Serbia 116

Jama'at-i-Islami 130
Janata Party (India) *see* Bharatiya Janata
  Party
Japan: and Argentina 72, 73; and
  Australia 89, 101; and Britain 41, 47;
  geopolitical visions 143; and Germany
  28; post-war views of 16; and
  Russia/Soviet Union 41, 101, 104,
  139; and USA 47, 54
Jaruzelski, Wojciech 31
Jerusalem 121
Jervis, Robert 61
Jews and Judaism 102, 117, 121;
  Zionism 57, 84, 121; *see also* Israel
Jezowa, Kazimiera 26–7
Jirinovsky (Russian politician) 101–2
John Paul, Pope 85
Johnson, Paul 158
Joyce, James 6
Judaism *see* Jews and Judaism
al-Jundi, Sami 161–2

Kaaba in Mecca 12, 120
Kafka, Franz 23
Kanan Makiya *see* Samir al-Khalil
Kapur, Harish 135
Karadžić (Bosnian-Serb President) 110,
  117
Kashmir 130, 132, 133, 134
Kates, R.W. 2
Kazakhstan 105
Keating, Paul 80, 89
Kedourie, E. 161–2
Kennedy, John F. 52
al-Khalil, Samir (Kanan Makiya) 8, 126
Khan, Ayub 133
Khouri, R.G. 161
Khrushchev, Nikita 14, 71
Kiev 100, 105, 106
Kipling, Rudyard 45
Kirgizstan 105
Kissinger, Henry A.: on age of transition
  146; and China 90–1; and
  'equilibrium' concept 62–3; on
  European balance of power 38;
  geopolitics described by 62; on lack of
  geopolitical tradition 4; and 'National

Bipartisan Commission on Central
America' 64, 65; on national debate
58; as new realist 59–60; and Nixon's
détente 63; on Vietnam 71
Kjellén, Rudolph 3
Koran 120
Korean War 132, 140
Kosovo, battle of 105, 112
Kozyrev (Russian Minister) 104
Krajna Serbs 115
Kundera, Milan 9
Kurds 11, 24, 117, 121, 126
Kuwait 144; invaded see Gulf War

Labour Party (Australia) 93
Lacoste, Yves 4
language: of foreign policy 5; political,
instability of 97
Lanús, Archibaldo 77, 83
Latin America 131; and USA 5, 63–5,
71; see also Argentina; Bolivia
Latvia 26
law and legal system 11, 62, 130
Leaver, Richard 92–3
Lebensraum 18, 20, 23, 28
legal system see law
Leland, Christopher T. 73
Lenin, Vladimir Ilyich 15
Lennox 85
Liberal Democratic Party (Russia) 101
Liberal Party (Australia) 87
liberalism: and Argentina 74, 76, 78;
and Russia/Soviet Union 95, 101
lines see borders
list of friendly and hostile nations 12
Liszt, Franz 25
literature 6; Argentina 76, 81, 82, 83;
Britain 36–8, 41, 44–5, 152;
geopolitical visions 144–5; Germany
20–2, 24, 76; USA 67–70;
utopianism in 53; see also fiction
local experience and identification: and
geopolitical visions 145, 146;
importance of 2, 3, 10, 16, 20, 135;
transcending national politics 7
longue durée 35
Loprete, C.A. 157
Lucas (film producer) 67
Luxembourg 10

McCutcheon, John T. 55
Macedonia 111
Mackinder, Sir Halford J.: on
Heartland-Rimland concept of
Eurasian core (expounded at RGS) 3,
14; and Argentina 72, 73; and Britain
40–1; and Germany 29; and India
128; and USA 60
McMahon line 133
Mahan, Alfred 3
Makiya, Kanan see al-Khalil
Malvinas (Falklands) War 73, 81, 82,
83, 84, 85
Manifest Destiny 52
Mao Zedong 7, 53
Marquand, David 47–8
Marx, Karl and Marxism 64, 96; and
Germany 15, 20–1, 24, 25; and India
131, 134; see also communism
materialism 24, 28; and idealism in
Germany 23–6, 28, 88; Russia/Soviet
Union 98
Matlock, Jack 95
May, Karl 20–2, 24, 76
Mayhew, Christopher 45–6
Mazowiecky Commission 109, 110
Mead, Walter R. 17, 18
Mecca: Kaaba 12, 120
media 1, 136, 143
Mediterranean area 39, 40, 75, 101,
106; see also Italy; Middle East; Spain
Meinvielle (priest) 84
Melanson, R.A. 63
Menem, Carlos 85
Mesopotamia 121–2, 126
Metropolis concept see city
Meynen, Richard 79
middle class: in Germany 22; in India
134, 135, 138; lacking in Serbia 115
Middle East: and Britain 39, 42, 44, 89;
and Russia/Soviet Union 105; see also
Egypt; Gulf War; Iran; Iraq; Kuwait;
Orientalism
migration see immigrants
Mihajlovic, Milenko 112, 113, 114, 116
Milan, King 115
military affairs: dictatorship in
Argentina 74, 75–6, 78, 84; see also
armed forces; occupation; wars
Milosević (Serbian President) 117
Ministry of Information (Britain) 43
missions and national identity 146–7
Mitteleuropa concept and central location
of Germany 14, 17–20, 24, 28–9, 33,
35, 144, 151
Mitterand, François 1, 15
Mladić (Bosnian-Serb general) 110, 117
models, other countries as 12, 20, 30–1,
93
modernization 14, 23–4, 96
Monroe Doctrine 65; as frontier 52;
Russian 103–6

Montenegro 109, 110, 112, 116
morality and family 57, 58
Moscow 15, 105; and Olympics 63; as
  third Rome 100, 101
Mosul 121
motivation, myths revealing 1
Moynihan, D.P. 134
Muhammad, Prophet 121
multilateralism 94
Munich appeasement 62
music and Germany 20–1, 24, 25
Muslims *see* Islam
Mussolini, Benito 79
'mutual understanding' 3
myths and symbolism 1–2, 11;
  Argentina 76, 81, 83–5, 89 (*see also*
  Péron); Australia 88–90, 93; of
  foreign policy aims not recognised 5;
  Germany 25–6; holy places 12, 99,
  120; invincibility of Muslim armies
  133; Iraq 119–20, 121, 122; order,
  restoration of 16; revealing
  motivation 1; Serbia 116; USA 49,
  52–3, 54, 56–8, 67, 70; *see also*
  borders; fiction; idealism; myths and
  symbolism, US

Naipaul, V.S. 73, 121, 127, 131, 135
Nandy, Ashis 137–8
Napoleon I, Emperor of France 1, 15, 37,
  38, 86, 99
national character: Australia 87–8;
  Serbia 110
national experience of place **1–16**,
  148–50; classical and modern
  geopolitics 3–6; genius of place 103;
  levels of identity and interpretation
  6–11; *see also* geopolitical visions
national identity: features of 11; *see also*
  'commonality'; countries; geopolitical
  visions; history; identity; myths and
  symbolism; national experience of
  place
national mission *see* collective mission
National Socialism 3, 18, 20, 22–3, 26,
  29, 34; and Argentina 73, 79, 80;
  and Britain 42, 46; and empty East
  28; *Geopolitik* 1; Nehru on 128; and
  Olympics 63; and Russia/Soviet Union
  98, 99; and USA 62, 71; and
  Wagner's music 21; *see also* Hitler;
  Second World War
NATO *see* North Atlantic Treaty
  Organization
'naturalness' of borders 12
Nazis *see* National Socialism

Nebuchadnezzar, King of Babylonia 121,
  122
negativity and identity 22, 89
Nehru, Jawaharlal 128–30, 131–2,
  133, 134, 135, 138
Netherlands: border 19; geopolitical
  visions 143; imperialism 28; low level
  of national pride 10; not complex like
  Germany 17; public opinion surveys
  143
neutrality and non-alignment 142;
  Argentina 80, 84; India 132, 134,
  138, 140
new realism: Australia 93–4; USA
  59–60
New Thinking 103–4
New York 10, 49
New Zealand 42, 86, 91, 92
Newman, J. 68
Nicholas I, Tsar of Russia 100
Nietzsche, Friedrich Wilhelm 34, 35
Nijinski, Vaslav 72
Nile, River 39
Nineveh 121
Nixon, Richard 60, 62, 63, 90, 155
non-alignment *see* neutrality and
  non-alignment
non-interference 135–7; *see also*
  isolationism
Non-Proliferation Treaty 134
non-violence 132
Noriega, Manuel 71
North America 21, 24; Canada 94, 98;
  see also United States
North Atlantic Treaty Organization: and
  Britain 44; and geopolitical visions
  142, 143; and Germany 20, 34; and
  Serbia 117; and USA 63
novels *see* fiction
nuclear reactor accident 32, 146
nuclear weapons: emergence of 42;
  bases in England, protest against 7;
  Cold War 3–4; fear of 144; India 134;
  Non-Proliferation Treaty 134; Russia
  69; survival after holocaust theme of
  US future wars fiction 68, 69
Nueva 85
Nuremberg trials 62

Ó Tuathail, G. (Gerard Toal) 5
Obst (German geographer) 29
occupation of Europe 9, 17, 46
oil 106
Oklahoma City explosion 49, 50–1,
  56–7
Olympics 63

Operation Desert Shield 91
Operation Desert Storm 63, 120
Operation Grand Slam 133
opinion surveys *see* public opinion surveys
order, visions and restoration of 15–16
organization theory 115
Orientalism 7–9
Orr, Carey 56
Orthodox Church 99–100, 102, 106
Ostforschung 27
Ostpolitik 30, 31, 32
Ostwachstum 19
other places *see* geopolitical visions; rest of the world
Ottoman Empire 39, 40, 100, 114–15
'Outer Crescent' 41

Pacific Ocean *see* South-east Asia and Pacific
Pahlavi, Mohammad Reza, Shah of Iran 123
Pakistan 129, 130; and China 137; and India 132, 133; and USA 133, 134
Palestine 121; Liberation Organization 92
Palmerston, Lord Henry 40
Panama 71; Canal 85
pan-Arabism 121
'partial commitment' 41
partisan resistance in war 116
party politics and elections: Argentina 74, 77; Australia 87, 93; Germany 143; India 130, 134, 135–7; Iraq 121; Japan 143; Russia/Soviet Union 101–2; USA 143; *see also* democracy
Patagonia 82
'peak-experience' 15; Cuban Revolution as 12–13; in Serbia 110, 116; *see also* Revolution; wars
Pellegrini, Carlos 72
perestroika 98
Perón, Juan/Peronism 78, 79–82, 85
Persia *see* Iran (Persia)
Peter the Great, Tsar of Russia 96, 100
Picton 85
place *see* geography; national experience of place
poetry 76; *see also* literature
Poland: and geopolitical visions 143; and Germany 26–7, 31, 32; and Russia/Soviet Union 39, 101, 104; seen as Oriental 8; and Serbia 114, 117; and USA 31
political correctness 33

political geography: France 4, 5; *see also* geopolitics
political parties *see* party politics
politicians and 'archetypes' 1
Popieluszko, Jerzy 31
Port, Mattijs van de 118
Porter, Bernard 39
post-modernism 7
post-Cold War ambiguity and USA 70–1
power: politics, hegemony as 64; sea 39; *see also* balance of power
pragmatism: Australia 87, 93–4; Britain 95; Germany 23, 24; India 137; *Realpolitik* 4, 23, 24, 137; USA 62, 95; *see also* realism
'prison of experience' 2
protectionism 47, 93–4
Protestants 37, 84
Prussia 26, 27, 73, 101
public opinion surveys 143–4; Australia 88; Britain 143; Cold War 43; Europe, attachment to 10; fluctuations 11; India 135–8, 143; Iraq 125, 127; Netherlands 143; Russia/Soviet Union 96, 98–9; USA 11, 64, 66–7, 143, 144; on Vietnam War 64, 66–7
Punjab 130–1
Puritans 59

Qadisiyya, battles of 119–20, 121

Racine, Jean-Luc 137
racism 93
Radical Party of Argentina (Unión Cívica) 74, 77
railways 40
Rao (Indian Prime Minister) 134
Rathenau, Walther 24
Ratzel, Friedrich 3, 8, 17, 18–20, 35, 151
Raum 27, 28
Rawlinson, Sir Henry 40
Reagan, Ronald 31, 60, 63–4, 65, 70, 71
realism: Australia 93–4; Germany 22, 26; India 129; reality and values 87; USA 59–60; *see also* new realism; pragmatism
Realpolitik see under pragmatism
Reformation 83
regeneration through violence 52
regional awareness 10
regionalism: Iraq 123–6; *see also* local
Reich: First 22, 28, 39, 41; *see also* Third Reich
Reichstag 19

religion: and identity in South Asia
130–1, 132, 133–6; *see also*
Christianity; God; Hinduism; Islam;
Sikhs
Renaissance 97, 100
renewal 58
republicanism in Australia 86
Republics, former Soviet 101, 103,
105–6, 107
residential changes as meaningful
movements 6
rest of world (and Us-and-Them attitude)
5, 7; and Argentina 72–3, 75–6,
78–84; and Australia 92, 93; and city
6, 10; and Eastern Europe 9; events
different in 2; fear of *see* dangers,
external *and* fear; and Germany 30;
and India 131–4; models, other
countries as 12; separation from *see*
isolationism; and USA 53–7, 59–61
(*see also* Vietnam War); *see also*
geopolitical visions; Orientalism
reunification of Germany: hopes 30; *see
also* 'Wende'
revenge 50, 110, 113; empire of *see*
Serbia
Revolution: of 1848 21, 22, 23;
Bolshevik 96, 100, 102, 140; Cuban
12–13; French 29, 84; as
'peak-experience' 12–13, 15; *see also*
wars
RGS *see* Royal Geographical Society
Rhine, River 10, 28–9
Richardson, James L. 91, 92
Ridley, Nicholas 12
rights *see* human rights
Rimland concept 155; *see also* Heartland
riots: in Argentina 79; in India 129–30;
in USA 49–50, 51, 52, 53, 56
rivers, importance of 8, 10, 28–9, 39,
100, 105
Roman Catholicism 8; and Argentina
79, 83–4; and Serbia 112, 114
Roman Empire 19
Rome 100, 101
Rosas, Juan Manuel de 75
Royal Geographical Society speech *see*
Mackinder
RSFSR *see* Russian Federation
Rumaila 123
Russia/Soviet Union **95–108**, 159–60;
and Africa 85; ambiguous future
95–9; and Argentina 79, 80; and Asia
15, 98, 101, 105, 106–8; and borders
97–8, 101; and Britain 39–40, 41,
43, 95, 105, 143; and China 4, 15,

63, 101, 106, 137; and Christianity
96, 98–101, 102, 105, 106; and city
100, 101, 105, 106; and Cold War
98, 99, 101; and Cuba 64; and
Eastern Europe 9, 14, 101, 106, 135,
143; and economy 95, 104; empire
foreseen 24; as enemy in US future
wars fiction 68–9, 70; and Eurasia
concept 15, 102, 106–8, 140, 144,
145; and Europe *see* under Europe;
and France 41; geopolitical reflex
absent 103–6; and geopolitical visions
139–46 *passim*; and Germany *see*
under Germany; and Gulf War 102;
and identity 99–102; and India 40,
98, 105, 128, 133, 134, 136; and
Iran 42, 44, 106; and Islam 99, 102,
106; and Japan 41, 101, 104, 139; as
Oriental country 8; and Poland 39,
101, 104; and public opinion surveys
96, 98–9; and Serbia 109, 114, 117,
118; and Turkey 106; and USA *see*
under United States; and World Wars
9, 98, 101, 102; dissolution of Soviet
Union 16, 57, 98, 144; *see also*
Afghanistan; communism
Russian Federation (RSFSR) 102, 103,
106, 107
Russo-Japanese War 41, 139
Rustam 119

'sacred geography' 129, 130
S'ad ibn Abi Waqqas 119, 120
Said, Edward W. 7, 8, 9
St Petersburg 15, 100, 105
Samir al-Khalil (Kanan Makiya) 126
SANU (Serbian Academy of Sciences and
Arts) 112
Sarajevo 116
Sarmiento, Domingo 75–6
'savages' and 'primitives', US romanti-
cization of 52
Savitsky, Pyotr 102
Scandinavia 98, 143
Schadenfreude 46
Schengen Pact 9
Schicksal 28
Schlesinger, Arthur M. Jnr 59
Schmidt, Helmut 17, 31, 35
Schmoller, Gustav 24
Schnass, F. 26
Schoultz, Lars 64, 65
science fiction: future wars 37, 67–70,
156
Scotland 107, 145
sea power 39

Second World War: and Argentina 80, 84; and Australia 89, 91, 92; and Britain 42, 45; casualties 61; and geopolitical visions 139, 141, 143, 144; and Germany 18, 22–3, 28, 30, 34, 35; and India 128; occupation of Europe 9, 17, 46; post-war period *see* Cold War; and Russia/Soviet Union (Great Patriotic War) 9, 98, 102; and Serbia 110, 112, 116; stability after 10; and USA 42, 54, 59, 61; *see also* Hitler; National Socialism
security 12, 144–6; risks 141, 163; *see also* Cold War; fear; wars
Security Council of UN 34, 138
Self and Other concept *see* rest of world
self-analysis in Argentina 81–3
self-centred small groups 58
self-confidence, excessive 123
self-criticism in Russia/Soviet Union 14
self-destruction 132
self-image: of elite in Russia/Soviet Union 14; as intermediary 141, 142
self-interest 93–4
self-realization 101
self-sufficiency 132; *see also* isolationism
Serbia 105, **109–18**, 160; and Britain 110; and geopolitical visions 140, 142; Greater Serbia 111, 112–17; history 109–11, 112; and Islam 109, 112, 114–15, 116; myth 116; and Poland 114, 117; royalty murdered 160; and Russia/Soviet Union 109, 114, 117, 118; and Turkey 110, 112, 113–15, 116; and World Wars 110, 111, 112, 116
Serbian Academy of Sciences and Arts (SANU) 112
Shatt-al-Arab 119
Shekhar, Chandra 136
Shiites 121, 126
Siberia 28, 40, 101, 106
Sicily 107
Sikhs 130, 131, 133, 136
similarities in world view *see* geopolitical visions
Singh, V.P. 136
Slavs 19, 27; *see also* Russia/Soviet Union; Ukraine
Sloan, G.R. 60, 63
Slotkin, Richard 52
Slovakia 101
Slovenia 111
Smith, Anthony 11
'social physics' 20
socialization of children 2

Solidarity 31, 32
Solzhenitsyn, Alexander 100–1, 102
Somalia 71, 102
Sondervolk, Dutch as 19
Sonderweg 33
South Africa 84–5, 137
South America *see* Latin America
South Asia *see* India; Pakistan
South Atlantic *see* Argentina; Malvinas
South Atlantic Treaty Organization 85
South-east Asia and Pacific 4, 40, 61; and Australia 86–7, 90, 91; *see also* Vietnam
sovereignty ideology of (Britain) 44–8
Soviet Union *see* Russia/Soviet Union
Spain: and Argentina 75, 76, 79, 81, 82, 83; borders 19; Civil War 79
Spanish-American War 53
Spiering, M. 37–8
Spies, August 50
Sprout, H. and M. 18
Spykman, Professor 128
Sri Lanka 130–1
Stalin, Joseph 14, 29, 43, 44, 71
state 145–6; Argentinian metamorphosis of 78–81; and Iraq 126–7; omitted from definition of geopolitical visions 11
stock market crash 27, 139
Stokes, Gale 115
strategic game, war as 69
Suez Canal 9, 44
Sumerians 121
Sunnis 121
superiority, British sense of 44–8
Swales, Martin 23
Sweden 143
symbolism *see* borders; myths and symbolism
Syria 126
system-opening 'coexistence' 29–32

Tacitus 19
Tadzhikistan 105
Tamil Nadu 130–1
Tamil Tigers 131
Taylor, A.J.P. 21
Taylor, Peter J. 41, 42, 43
territorial borders *see* borders
terrorism and anarchism, fear of: Argentina 77; USA 49–51, 52, 53, 56–7
Teutonic Order 25–6
Thatcher, Margaret 12, 38, 46
'theatres of military activities' 105
Them-and-Us *see* Us-and-Them

Thies, Jochen 17
Third Reich 31; *see also* National Socialism
Third World: and geopolitical visions 139, 143, 145; nationalism 4; and Russia/Soviet Union 104; and USA 63, 65 (*see also* Vietnam); *see also* Africa; Asia; India; Latin America
Third World War, fear of 63; *see also* Cold War; fear of communism
Tiananmen Square 131, 136–7
Tibet 133
Tito (Josip Broz) 111
Tokyo trials 62
trade *see* economy and trade
Trans-Siberian Railway 40
Treitschke, Heinrich von 20, 25–6
Trinidad 131
Trubetskoy, Prince Nikolay 102
Truman, Harry S. 44, 63; Truman Doctrine 60–1
Turkey: and Britain 39, 40, 42; and Iraq 127; Ottoman Empire 39, 40, 42, 60, 106, 114–15; and Russia/Soviet Union 106; and Serbia 110, 112, 113–15, 116; and USA 60
Turkmenistan 105
Turner, Frederick J. 52, 53
TVDs ('theatres of military activities') 105
two-fold division of world and isolated countries (USA, Argentina and Australia) 141; *see also* Cold War; East and West

Ukraine 100, 101, 105, 106, 108
unaligned position *see* neutrality
Un-American Activities *see* witch-hunt
unconscious standards applied 3
understanding, 'mutual' 3
United Nations: and Britain 34, 44; and Russia/Soviet Union 102; and South Asia 132, 135, 138; and USA 50
United States **49–71**, 153–6; American way of life 5; and Argentina 72, 73, 80, 83, 85; and Australia 88, 89, 90–4; bases in England, protest against 7; and borders 1–2, 49, 51–3, 55–7, 67; and Britain 7, 28, 38, 42–4, 46, 84, 140; and Central America 5, 31, 63–5, 71; and China 60, 90–1; and Cold War 5, 16, 54–6, 57, 62; and communism *see* Cold War, Vietnam War *and* witch-hunt; discontinuities in global perception 59–61; empire foreseen 24; and

Europe 50–7, 59; explosions compared 49–51, 52, 53, 56–7; fear of Russia/Soviet Union 16; frontiers *see* borders *above*; and geopolitical visions 139, 142, 143–4, 145; and Germany 17, 31, 32, 33, 34; and Gulf War 70–1, 73; and idealism 2, 57, 88; and identity 51–3, 140; immigrants 50, 51, 59; and India 128, 134, 136, 137; and Iran 60, 63; and Iraq 120, 121, 125; isolationism 51–2, 53, 54, 59–60; and Japan 47, 54; as leader *see* Cold War; as model 12, 20; myths and symbolism 49, 52–3, 54, 56–8, 67, 70; occupation of Europe 9; and Pakistan 133, 134; and Poland 31; post-Cold War ambiguity 70–1; and public opinion 11, 64, 66–7, 143, 144; purity ideal 2, 57; and rest of world 53–7, 59–61 (*see also* Vietnam War) ; and Russia/Soviet Union 43, 60–1, 95–6, 101, 102, 106 (*see also* Cold War); security policy 12; and Serbia 110; and Third World 63, 65 (*see also* Vietnam War); and Turkey 60; and wars 42, 51, 53–4, 57, 59, 61 (*see also* communism *above*)
unity and diversity in India 128–31
universalism 7
Unsworth, M. 68
Urban, Michael 97
Us-and-Them attitude *see* geopolitical visions; rest of world
Ustasha 112, 114
utilitarianism 93
utopianism in literature 53
Uzbekistan 105

Varenne, Hervé 9, 10
Varshney, Ashutosh 129
Versailles, Treaty of 27, 54
victimization 57–8
victorious America theme of US future wars fiction 68, 69, 70
Vidmar, Josip 111
Vietnam War and USA 5, 59, 63, 70, 71, 117; and Australia 90; casualties 61, 62; and geopolitical visions 140, 141, 143; Green Berets 52–3; post-War faith in intervention 63–4; post-war syndrome 64; spatial pattern of bombing 4; withdrawal 60
Villegas, General 85
violence: eschewed *see* neutrality; in US

myth 49, 52, 56–7; *see also* invasion; Revolution; riots; wars
Vistula, River 28–9
Vojvodina 115
Volga, River 105
völkisch 22
vulnerability, zone of 39

Waco tragedy 57
Wagner, Cosima 25
Wagner, Richard 20–1, 24, 25
Waisman, Carlos H. 78, 80
Walker, Martin 98
Wallace, W. 46, 47
wars: Argentina: between state and citizen 75, 76–7; Argentina: Indians 75, 76; Balkan (1912, 1913) 110; Balkan since dissolution of Yugoslavia *see* Serbia; and Britain 37, 38, 42, 45; casualties 61–2, 71, 89; Crimean 40; crimes 62; future fictional 37, 67–70, 156; and geopolitical visions 144; guerilla 12–14, 52; Indo-Chinese 133, 140; Indo-Pakistan 132, 133; Iran-Iraq 120, 122, 126, 141; Korean 132, 140; military meaning of geographical data 4; military occupation of Germany 9, 17, 46; Russo-Japanese 41, 139; Spanish-American 53; as 'sublime events' 25; unnecessary 39; World *see* First World War *and* Second World War; *see also* armed forces; Civil Wars; Gulf War; invasion; Revolution; security; Vietnam War
Warsaw Pact countries *see* Eastern Europe
Watergate 64
weapons *see* nuclear weapons
Wehner, Herbert 31

Weichsel, River *see* Vistula
Weimar Republic 22
'*Wende*' and end of Berlin Wall 1, 30; after 32–5; US reactions to 17
Wendt (historian) 18
White Australia policy 93, 139
Whitlam, Gough 90–1, 92
William IV, King of Britain 115
Wills, Gary 51
Wilson, Sir Robert 39
Wilson, Thomas Woodrow 54, 58
witch-hunt against communism in USA 5, 42, 50, 54, 58, 139
Wittkopf, E.R. 67
Wolpert, Stanley 135
world balance of power *see* balance of power
world empire concept 72; *see also* Heartland
world order, yearning for 151–6
'world peace' 7
world view, similarities *see* geopolitical visions
World Wars *see* First World War; Second World War
Wynia, Gary 78, 81

Xinhua press agency 137

Yeltsin, Boris 134
youth concept in USA 52, 58
Yrigoyén (Argentinian President) 74
Yugoslavia, former 3, 109–10, 111, 115, 140, 142; *see also* Bosnia; Croatia; Serbia

zero-sum view of international relations 5
Zharikhin (Russian Civic Union leader) 97
Zionism *see under* Jews and Judaism